MW00509496

Airplanes
and
Safety

"*Soon shall thy arm, unconquered Steam, afar
Draw the slow barge, or drive the rapid car;
Or, on wide waving wings expanded, bear
The flying chariot through the field of air.*"

Erasmus Darwin. [1781]

THE TRAVELERS
HARTFORD, CONNECTICUT

22031 4-13-21

PREFACE

*B*EFORE *aircraft can be extensively utilized for private and commercial purposes, and before aerial navigation can be developed to a point where it will afford an attractive field to insurance companies, it will be necessary to effect a substantial readjustment of present-day conditions. The public, for example, will have to acquire a considerable amount of aeronautical knowledge before it will be prepared to admit the practicability of aerial navigation. There is also a crying need for a vastly greater number of official landing fields, laid out and managed in accordance with approved safety principles. It is likewise necessary to establish standard airways, duly provided with aerial lighthouses and wireless signal stations; and to pass uniform and stringent laws governing the licensing of pilots, the construction and use of aircraft, and the conduct of air-navigation generally.*

*In the present stage of development it is impossible to discuss the subject of "*AIRPLANES AND SAFETY*" exhaustively and fully, and the present book does not attempt to do so. We are putting it forth, however, in the* *belief that it will assist in promoting aerial navigation by presenting an elementary account of the construction and operation of airplanes, and by discussing some of the means by which flying may be made a safer mode of travel and transportation.*

Rapid changes of many kinds are inevitable in connection with so new a subject,—changes relating not only to apparatus, but also to the training and licensing of pilots, to the insurance coverage, to the legal aspect, and to many other phases. With increasing study and experience the subject is sure to develop quickly, and it is quite within the range of possibility that fundamental modifications along any of these lines may be forthcoming within a year or two. - If we had the gift of prophecy we should include all these future advances and improvements in the present book. Not having any such gift, however, we have merely endeavored to represent the subject as it stands to-day.

The text is based primarily upon our own experience and observation, though we have naturally consulted numerous books, pamphlets, reports, and technical journals. Furthermore, we have had friendly personal counsel from many eminent and qualified sources, and notably from Colonel E. A. Deeds of the Equipment Division, Air Service, U. S. Army; from Lieut.-Colonel H. M. Hickam, Major E. L. Jones, and L. D. Seymour, M. E., of the Information Group, Office of the Director of Air Service; from Lieut.-Commander Byrd and Mr. Lane Lucy, of the Navy Department; from Second Assistant Postmaster General Otto Prager; and from Nelson S. Hopkins, President of the Phenix Aircraft Products Company. But although we desire to express our fullest appreciation of the generous assistance we have received from these gentlemen, it must be clearly understood that they have acted only in an advisory capacity, and that they are not in any way responsible for the statements that are made. The responsibility is wholly our own, and we accept it in full.

THE TRAVELERS INSURANCE COMPANY
THE TRAVELERS INDEMNITY COMPANY
HARTFORD, CONNECTICUT

CONTENTS

ILLUSTRATIONS

Airplanes
and
Safety

INTRODUCTION

THE LEGEND OF ICARUS: Men dreamed of
navigating the air long before any means were
devised for realizing such dreams. Legends, some of
which run back into remote antiquity, tell of men who
succeeded in rising into the air by one means or another,
—the best known story of this kind being the one in
which Daedalus and Icarus figure. It may be remembered that Daedalus, who was a talented Greek inventor,
fled for certain reasons to the Island of Crete, where he
constructed a famous labyrinth for King Minos, pre-
sumably about the year 2000 B. C. Subsequently,
Daedalus was himself imprisoned in this labyrinth,
together with his son Icarus; and we are told that they
saved up the feathers that fell into their prison from
birds passing overhead, and eventually fashioned these
feathers into wings, by means of which they effected
their escape. Daedalus told Icarus to keep up high
enough to avoid the dampness from the sea, but
warned him not to fly too near the sun lest the heat from
it melt the wax by which the feathers in his wings were
held together. The young man disregarded the latter

part of this counsel, however, and the accident that his father had foreseen came to pass. He flew too high, the wax softened, the wings became unmanageable, and the youthful aviator fell into the sea and was drowned. Tradition even records the place of his fall, locating it near the island of Samos; and the part of the Aegean Sea in that vicinity is still called the Icarian Sea. According to this evidence, the young man must have flown something like 200 miles before he came to grief. The father escaped in safety, and eventually reached Sicily.

Every legend probably has some measure of foundation in fact, but in the case of Icarus it would be hard to identify the elements of reality, and separate them from the frills and wrinkles that forty centuries have added. The labyrinth to which the story refers has recently been discovered and explored, and we also know that Crete, in the time of Minos and Daedalus, was one of the world's most influential and important centers of progress and civilization. All this, however, proves nothing about the reality of the overseas flight of Icarus. The most realistic and probable element of the tale, from our point of view, is the disregard that Icarus showed for the safety advice that his father gave him. The older man was an experienced mechanic and inventor, who realized that accidents are likely to happen, and who was thoughtful enough to consider, in advance, how safety could best be assured. But the younger man, just like millions of others down to the present day, let the counsel pass in at one ear and out at the other; and when he was soaring up into the sky he "took a chance," and almost immediately thereafter he was killed. That doesn't sound the least bit like forty centuries ago. It sounds more like last Wednesday afternoon, in the little shop across the street.

Mr. Will H. Low has painted a beautiful panel in the New York State Education Building, depicting the fall of Icarus; and by special permission of the artist and the New York State Education Department we have used a photo-engraving of it as the frontispiece of this book. The modern airplane, high up in the sky of the painting, typifies the success that has finally followed in the wake of so many years of dreaming and experimenting. The body of Icarus lies upon the rocky shore where it has been cast up by the sea,— the nearness of which is suggested, or symbolized, by the water in the immediate foreground.

EARLY BALLOONS: Passing, now, from legend to history, we find that the first successful attempt to navigate the air was made in France, by two brothers, Stephen and Joseph Montgolfier. Toward the end of 1782 they had found that light bags would ascend if filled with heated air, and on June 5, 1783, they gave a public exhibition near Lyons, in the course of which a linen globe more than thirty feet in diameter was inflated with hot air and liberated. It rose to a great height, remained in the air about ten minutes, and came down again about a mile and a half from the starting point. This experiment naturally attracted a great deal of attention, and the French physicist J. A. C. Charles, realizing that a hot air balloon would be impracticable except for extremely short flights, suggested that the necessary levity be obtained by filling the bag with hydrogen gas (which was then called "inflammable air"). The money required for testing the practicability of the idea was raised by popular subscription; and on August 27, 1783, a gas-filled balloon, thirteen feet in diameter, constructed by the Robert brothers under the direction of Charles, ascended from the Champ-de-Mars, Paris. It rose to a height of 3000 feet and remained in the air for three quarters of an hour.

On September 19, 1783, Joseph Montgolfier sent up another hot-air balloon at Versailles in the presence of the King and Queen and a vast assemblage of other spectators, and on this occasion a cage was taken up, containing a sheep, a rooster, and a duck. The rooster and the duck behaved themselves with due dignity, but the sheep, failing to understand that it was about to become famous as a member of the first party of living creatures to go up in a balloon, or not appreciating the honor thus thrust upon it, kicked the rooster just before the start, and injured it somewhat. The ascent was made successfully, however, and without further harm to the creatures in the cage, although they ascended to a height of about 1500 feet and came down two miles from the starting point, after a trip lasting eight minutes.

The practicability of making a balloon ascent being thus demonstrated, men were soon found who were willing to take the attendant risks. The first human being to ascend was Jean François Pilâtre de Rozier, who went up in a captive balloon on October 15, 1783, with a lighted brazier suspended below the bag to keep the air heated. After a number of experiences of this kind he ventured to go up in a free fire balloon, on November 21, 1783, in company with the Marquis d'Arlandes. They remained in the air more than twenty minutes, and came down safely after drifting more than five miles.

Ten days later (namely, on December 1) Charles, the physicist whom we have already mentioned, ascended from Paris, accompanied by one of the Robert brothers, in a balloon filled with hydrogen. The two men were in the air about two hours, and landed some twenty-seven miles from the starting point. Robert then left the car, and Charles made a second ascent alone, rising on this occasion to a height of about two miles.

Honorable mention must be made, at this point, of Rittenhouse and Hopkinson, of Philadelphia, who were experimenting with balloons almost as early as the Montgolfiers, and who also constructed a successful gas-filled balloon of the composite type, in which James Wilcox made an ascent.

Girond de la Villette, who had accompanied de Rozier in one of his early ascents, proposed to employ balloons in warfare; and in 1794 a French company of "aerostiers" was formed, an air-park was established, and balloon reconnaissances were actually made against the Austrians. Balloons were used extensively for observation work during the siege of Paris, in 1870-71. They were also used in considerable numbers in our own War between the States, and in the Spanish War of 1898, and the Russo-Japanese War. All these military balloons were of the spherical type, and were moored to the ground by means of cables. The modern stream-lined observation balloon was not developed until a few years prior to 1914, but before the close of the World War, it had practically displaced all other means of directing artillery fire.

THE BEGINNINGS of the Airplane: The first steps towards the development of the heavier-than-air flying machine—that is, of the type of aircraft that has no supporting gas-bag—may be assigned to various periods, according to the views that we may hold with regard to what constitutes a first step. If we chose to go back to the most elementary principles of mechanical flight, in which other elements than mere momentum are utilized for keeping the apparatus in the air, it would doubtless be necessary to credit the prehistoric aborigines of Australia with the earliest invention, because the boomerang certainly involves some of the principles that underlie the operation of the modern airplane. If, on the other hand, we are to pass over the boomerang, as well as the kite (including the scientific species as well as the juvenile one), and the various small-sized "helicopters" and other toy-like devices that have appeared from time to time, and are to begin our survey with the earliest form of apparatus that held out distinct promise of an actual, early solution of the problem of navigating the air with a heavier-than-air mechanism, we shall have come down to the latter part of the Nineteenth Century, when Lilienthal, Chanute, Pilcher, the Wright brothers, and many others, laid the foundation, by means of their gliding planes, for the rapid development of the art in a useful and practical direction.

Dr. Samuel Pierpont Langley, Secretary of the Smithsonian Institution, will always be regarded as the first to establish the principles of mechanical flight upon a sound scientific basis. He began his serious work along this line about the year 1887, publishing his *"Experiments in Aerodynamics"* in 1891, and *"The Internal Work of the Wind"* in 1894.

Sir Hiram Maxim constructed an enormous steam-driven airplane in 1894, for experimental purposes. It was supposed to be confined to a long track that was erected for the purpose, but it tore loose, left the track, wrecked itself, and was never rebuilt.

Langley built a quarter-size steam-driven airplane which made a sustained flight over the Potomac river near Washington, D. C., on May 6, 1896. Encouraged by this result, he afterward constructed a full-sized machine, also driven by steam, and flights were attempted near Washington on October 7, 1903, and again on December 8. Owing to mishaps in launching, the machine fell into the Potomac on both occasions, and on the second trial it was wrecked by the poorly-directed efforts of a tug to rescue it. The failure of the experiment was witnessed by hundreds of newspaper correspondents, who published supposedly humorous accounts of the proceedings, and the ridicule caused Congress to refuse further contributions toward the work, and also had a profound effect upon Langley himself, so that many believe that it hastened his death. The machine was raised and placed on exhibition in the Smithsonian Institution. In recent years it has been repaired, and after being provided with new wings and with a gasoline engine, it has been successfully flown,—the soundness of Langley's general design being thereby proved.

In 1903 Wilbur and Orville Wright built a self-propelled machine in which they used a twelve-horse-power gasoline engine and two propellers. Their first successful flight with this design was made on December 17, 1903. In 1904 and 1905 they made many flights, some in public and some in private. They

Photo by Benner.

THE RECONSTRUCTED LANGLEY "AERODROME."

originally undertook their aerial experiments out of
pure scientific interest, and with no thought of a
possible commercial return. As their investigations
proceeded, however, they became so absorbed in the
subject that they gave up all other business and devoted themselves solely to their aircraft researches.
In the winter of 1907-8 the U. S. Signal Corps called
for bids on an airplane and an airship, and the Wright
brothers undertook to comply with the conditions and
deliver a practical airplane. With this in view they
did a great amount of experimental work during 1908,
and in the fall of that year they began, at Washington,
a series of demonstration flights which terminated in the
unfortunate death of Lieut. Selfridge, who was a passenger with Orville Wright. Delivery of the machine to
the United States Government was finally made in 1909.

The pioneer flights made by the Wright brothers
in 1904 and 1905 were followed in 1908 by the work of
the Aerial Experiment Association, composed of Dr.

A. G. Bell, Glenn Curtiss, Lieut. Thomas E. Selfridge, F. W. Baldwin, and J. A. D. McCurdy, and by the flights of Curtiss in his own machine in 1909.

The achievements of Langley, the Wright brothers, and Curtiss, gave to the United States the distinction of being unquestionably the first country in the world to build what were conceded to be successful airplanes.

INFLUENCE of the World War: From the beginnings here outlined, progress was slow. At the outset there was practically no demand for airplanes, and the few that were used for sport were of a primitive type. The United States Army did a little flying, but no serious attempt was made to develop this branch of the service until 1914. Five officers of the United States Army were then sent to the Massachusetts Institute of Technology to study aeronautics, and in August, 1914, these five men constituted the entire technically-trained personnel of our Army air service, though there were, in all, twenty-four officers and one hundred and fifteen enlisted men on aeronautical duty on that date.

The problem that confronted the United States in connection with aerial navigation at the time we entered the World War was a staggering one. With no stock of material, and with practically no personnel experienced in airplane designing, and with an utter lack of knowledge of the requirements of the most advanced aircraft for war purposes, or of the appliances essential to their operation, the government faced a serious situation. But the war brought forth marvelous unrealized resources, both of materials and of technical knowledge, and in spite of contradictory opinion, the

development of our Air Service reflects the greatest
credit upon the men who handled the situation; and
if the war had continued six months longer, the United
States would have supplied more airplanes than all
of our Allies combined. At the time of the armistice,
some of our heavy planes were already superior in design to those of European make.

By the co-operation of the manufacturers in this
country, and with the assistance of our European
Allies, flying fields and training schools were developed
and engines and planes were perfected with great
rapidity; and at the end of the World War our air
forces numbered nearly 200,000, including 20,708
trained officers and 174,456 enlisted men and civilian
employees. Twenty-seven flying fields were then in
operation, and 9,503 training airplanes and 642 observation balloons of various sorts had been built. In addition to this, 17,673 aeronautical engines for training
purposes had been completed, and the work was still in
a state of rapid development when it was stopped by
the ending of the war. Considerable money was
doubtless spent without commensurate material re-
turns, but this was unavoidable in view of the nature
of the problems that had to be handled, and the pressure under which the work was done. The material
output was immense, however, and highly creditable
under the circumstances. In addition, our knowledge
of aeronautics was vastly increased, aeronautical engineers were developed, new methods of doing work
were evolved, and special materials were devised for
fulfilling special needs. During this period the advance
was so rapid, in fact, that airplanes sometimes became
obsolete almost before they could be completed.

THE FUTURE of Aerial Navigation: With the close of the war, it became necessary to consider the future of aeronautics.   Many persons believed that aircraft would be found to be useful in connection with the arts of peace, and the success of aerial navigators in crossing the Atlantic Ocean certainly took the question of the commercial possibilities of aircraft out of the province of the dreamer, and forced the practical business man to give serious consideration to the subject.   Yet the difficulties to be overcome are undeniably great, and many of our best engineers have been profoundly skeptical with regard to every suggestion involving the use of aircraft as a means of transportation in time of peace; and the general public, remembering the numerous accidents that have been recorded in our newspapers during the past few years, is strongly disposed to question the feasibility of this mode of travel.   But it should not be forgotten that the experience during the war is not a fair index of what can be accomplished in the future,—not only because the conditions existing at that time were far from normal, but also because the entire art was wholly new, and involved difficulties that are only now coming to be fully understood.

COMMERCIAL USES of Aircraft: European countries took to commercializing aircraft before the United States gave much thought to the subject, and several regular aerial lines of travel have already been established over there.   In our own country, an aerial mail service has been in operation on certain routes since 1918.   Experience shows this service to be practical and worthy of further development in the future. New routes are being established, and the size of the

Division of Aerial Mail Service is rapidly increasing.

Aircraft have also been employed, for a considerable time, by the Forest Service in fire-patrol duty. The forests of California from San Francisco to the Mexican border are regularly patrolled in this way, and the record established has been excellent. Observation balloons are used as stationary outlooks, and airplanes are employed to cover specified routes daily.

The practicability of using aircraft for photographic survey purposes, and to some extent for mercantile delivery and passenger service, has been demonstrated beyond a doubt; and if commercial aviation receives the necessary financial support, it will probably be only a short time before this means of rapid transportation will be established on a sound business basis;—provided a sufficient number of suitable landing fields are established, and adequate flying laws and regulations are enacted and enforced.

I. AIRPLANE CONSTRUCTION

TYPES OF MACHINES: Self-propelled aircraft may be divided into two main classes, according as they are (1) lighter than air, or (2) heavier than air. Lighter-than-air machines (technically known as airships or dirigibles when they are provided with engines and propellers so that they are capable of independent locomotion) are of the balloon type, and owe their lifting power largely or wholly to bags filled with a gas that is lighter than air. Such machines may be divided into (1) rigid, (2) semi-rigid, and (3) non-rigid types. In rigid airships the gas envelopes are supported by a rigid framework. The semi-rigid airships have a framework to support the cars, fins, rudders, and elevators, and non-rigid types owe their firmness entirely to the pressure in the gas envelope. The heavier-than-air machine is of totally different construction and owes its lifting power to the action of wings, or to the rotation of propellers analogous to the screw propellers that are used on steamships. If the machine were supported by wings that flapped like those of a bird, it would be called an "ornithopter;"

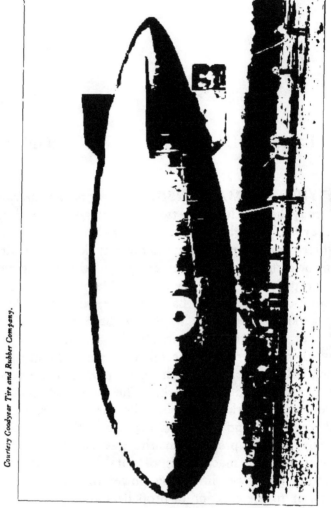

Courtesy Goodyear Tire and Rubber Company.

A NON-RIGID DIRIGIBLE OR BLIMP.

and if it were sustained in the air by the direct thrust of downwardly-directed propellers, it would be called a "helicopter." Neither of these types has yet been developed to a practical point in connection with gasless machines, although some authorities believe that the helicopter will become practicable in the near future. At the present time, all gasless machines of the "airplane" type owe their lifting power to the action of wings or supporting surfaces which are fixed and practically rigid, save for the fact that certain relatively small portions of them can be set or adjusted in varying positions. The fixed wings or "planes" are designed and located so that when the machine is in rapid forward motion, the air presses against their lower surfaces and also produces a vacuum over the top of the wing, behind the leading edge. It is usually considered that about 60 per cent. of the lift is due to the presence of this vacuum over the upper surfaces of the planes.

The wings were originally made thin and nearly flat, and it was then appropriate to call them "planes." In recent machines the wings are often quite thick, and they invariably have a strongly-curved shape also. It is hardly appropriate, therefore, to call them "planes" at the present day, though the name still persists, and the machine itself will doubtless always be known as an "airplane."

Airplanes (to which our attention will be almost wholly confined in the remaining part of this book) may be classified in various ways. First, they may be grouped in accordance with the number of main supporting surfaces employed,—"monoplanes" using one pair of such surfaces, "biplanes" using two, and "multiplanes" using more than two. They are also classified

Photo by Benner.

An Airplane.

Photo by Benner.

A HYDROAIRPLANE.

according to the nature of the service for which they are
designed, being known simply as "airplanes" if they are
to operate exclusively from the land, and as "seaplanes"
if they are to be used for marine flying. Seaplanes
are further divided into float seaplanes or "hydroair-
planes," and boat seaplanes or "flying boats." Hydro-
airplanes are similar to ordinary airplanes in construction,
except that instead of having landing carriages and
wheels, each machine is provided with a float or a set of
floats, for landing purposes. A flying boat is a seaplane
in which the body of the machine acts as the float.

Photo by Benner.

A FLYING BOAT.

Aeronautical Engines: Aeronautical engines are of the internal-combustion type, and use gasoline as fuel. They are all multicylinder in design, and may be either rotative or fixed. Fixed engines may be further subdivided, according to the arrangement of the cylinders, into radial, upright, and V-shaped types. The number of engines carried by an airplane varies, —some planes employing but one each, while others have two, and some designs call for three or more.

In the matter of locomotion, airplanes are usually either "tractors" (in which the propellers are located in front of the engines and *pull* the machines through the air) or "pushers" (in which the propellers are in the rear of the engines, and force the airplanes forward by a pushing action). In a pusher plane, the motor is usually mounted above the body; while in a tractor airplane it is commonly placed at the nose of the fuselage. Some multimotored types, however, combine the pusher and tractor principles in a single machine.

The Structural Parts of Airplanes: Excluding the power plant, an airplane can be divided into four principal parts: (1) body; (2) wings; (3) tail; and (4) landing-gear. We proceed to describe these briefly, as they are constituted in airplanes of the usual types. In all-metal planes, which will be mentioned later, the construction of the various parts is quite different.

The Body: The body of a tractor plane is termed the "fuselage," and in the pusher-type of machine it is shorter and is called the "nacelle." The fuselage or nacelle of the airplane usually carries the dead weight, consisting of the power plant, the fuel and oil tanks, and the pilot, passengers, and freight. It must be strongly built, and be constructed so that it will easily and safely transmit the forward pull or thrust from the propeller to the rest of the machine. This requires rigid attachment between the body and the wings. In flying boats the body construction is heavier than in other airplanes, because in this type the body serves also as a large landing pontoon, and keeps the plane afloat when it is resting on the surface of the water.

The Wings: The main supporting surfaces of an airplane are made up of several parts,—namely, wing spars, wing ribs, and the wing covering;—and wires and cables are used for internal bracing. The wing spars run longitudinally along the wings, receiving the stress to which the wings are subjected, and transmitting it to the framework of the body. There are usually two of these spars in each wing, one being located near the leading edge, while the other is ordinarily placed at about one-fourth of the length of the wing chord from the trailing edge. The wing spars are usually made of wood, and they may be either solid or built up of several pieces glued together. If solid, they are ordinarily made of ash or silver spruce; and if built up, they are made in differing combinations, varying from plain strip plywood to "I", "U", and box-shaped sections.

Wing ribs are employed to give the wing its shape and to complete the framework over which the wing covering is stretched. They are fitted transversely between the wing spars, and specially designed ones take the compression between these front and rear members. The ribs may be either solid or built-up; and in airplanes of some types special lightly-con-structed intermediate ribs are used for maintaining the shape of the wings, especially in the nose of the wings.

The wing covering, and sometimes the covering for the body frame, consists of linen or cotton cloth drawn tightly over the framework and sewed in place. This fabric covering must be perfectly smooth and taut. When in position, it is coated with a special varnish-like preparation, technically called "dope," which shrinks the cloth and at the same time makes it waterproof. The warp and filling threads of the wing cloth should be of uniform thickness throughout their length, and they should be as long as possible, because knots are likely to cause weak, uneven spots in the cloth. Some kinds of wing cloth, however, have extra stout threads (known as guide threads) woven at intervals in both the warp and filling, for the purpose of preventing the extension of any split or flaw that may develop in the fabric. Linen is much stronger than cotton cloth of the same weight, and it also takes the dope better. It should be used when its cost, per yard, is not more than 50 per cent. greater than that of special airplane cotton.

In the assembled biplane or multiplane, wires, cables, and interplane struts are used to form a truss-work between the main supporting surfaces. The interplane struts of a·biplane or multiplane machine are all in compression, and (like the wing spars) they may be either solid or built up. They serve to keep the wings at the proper distance apart, and in most planes they are placed vertically, or nearly so, between the wings. Struts are usually made of spruce or ash, but steel struts are now used to some extent, and are giving excellent results. The struts are attached to the wing spars by a metal-socket arrangement, and it is best to fasten the socket to the spar by means of a U-shaped bolt which passes around the spar instead of piercing it.

The wiring in the wings of an airplane is one of the most important factors in the construction of the machine. For practical purposes, the wires may be divided, according to their uses, into four classes,— flying, landing, drag, and incidence wires.

Flying wires are used to support the weight of the fuselage or body, when the machine is in the air. They extend from the bases of the struts on the lower wings, upward and *outward* to the tops of the struts on the upper wings. The landing wires pass from the base of each strut on the lower wings, upward and *inward* to the top of the next strut. They cross the flying wires nearly at right angles, and are under tension only while the airplane is on the ground.

When a machine is flying, the resistance of the air has a tendency to push the wings backward. This drift-back is counteracted by the drag wires, which, in a tractor machine, are usually attached to the front of the fuselage and extend back to the lower ends of the

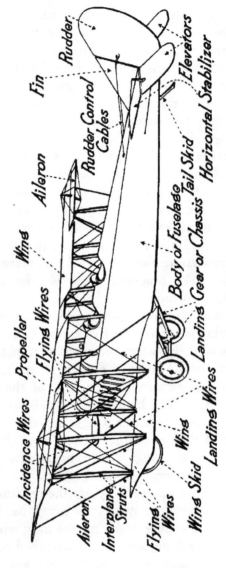

DIAGRAM SHOWING THE PARTS OF AN AIRPLANE.

struts. In a pusher-type machine the wires are usually attached to the front of the outrigging, or to the nacelle, and extend to the outer struts.

Incidence wires are used to adjust and maintain the angle of incidence. They take the form of cross-bracing between each front strut and the corresponding rear strut. Incidence wires are sometimes called "stagger-wires," because they are also used in regulating the stagger of the wings.

As a rule, all wires are adjustable by means of turn-buckles. This arrangement makes it possible to slacken or tighten the wires when necessary. The threads on the turnbuckles should be strong enough to insure safety, and a locking device of some kind should be used on them, to avoid any possible chance of the adjustment becoming changed accidentally, or in consequence of the vibration of the machine.

Modern airplanes have "ailerons," or hinged flaps, attached to the rear edges of the wings, at or near their outer ends. These are used to stabilize the machine laterally. As long as the ailerons remain in a neutral position they do not disturb the equilibrium; but as soon as the ailerons on either side are raised or lowered, the machine tends to tilt sidewise;—that is, to rotate one way or the other about its fore-and-aft axis. They permit the pilot to right his machine in case it should be tilted by a gust of wind, and to tilt or "bank" the machine when a turn is made. Ailerons that are long and narrow are said to be more effective and more easily operated than short ones having the same area. The aileron surface should be equal to about 10 per cent. of the wing area.

The Tail: There are two distinct types of tails,

—the lifting and non-lifting. The lifting tail may have a surface cambered similarly to the main supporting surface, but in most cases the angle of incidence at which the tail planes are set causes a slight lift. A lifting tail supports only its own weight however, and contributes nothing to the support of other parts of the machine. The non-lifting tail has a surface either flat or possessing a slightly convex camber on both sides. It is designed to act as a horizontal fin or stabilizer, and is set so as not to have a lifting effect, but to give steadiness in a fore-and-aft direction, and prevent motions of the airplane analogous to the "pitching" of a ship.

The trailing edge of the horizontal stabilizer is usually constructed of metal tubing to form the rear spar, and the elevator planes are hinged to this spar. The stabilizers and elevators should have a combined area equal to about 15 per cent. of the total wing surface. Elevators are said to be more effective if built long and narrow.

A vertical stabilizer or fin is placed along the upper center-line of the tail, and extends back to the tail post of the body. The rudder is attached to the trailing edge of this fin, and to the tail-post, by hinges and pins, in a manner similar to that employed in attaching a rudder to a ship. The rudders and fins should have a combined area equal to about one-half that of the stabilizers, and should be located so that they will not be blanketed by the airplane body.

The Landing Gear: The chassis of an airplane is usually a V-shaped structure, strongly cross-braced and fitted to the lower side of the fuselage. Similar cross-braced construction is employed on seaplanes. The landing gear or undercarriage has two distinct forces to resist: (1) the vertical shock experienced upon landing, and (2) the horizontal force that tends to sweep the landing gear backward when the airplane is running along the ground. The former of these forces is greatest when a machine is "pancaked," and the latter reaches its maximum when a fast landing is made on soft or rough ground. A large factor in the landing stress of an airplane is what is commonly known as "side-swipe". This is partially counteracted by means of cross bracing in the landing gear.

The vertical shock is relieved to some extent by the use of rubber shock-absorbers, and the horizontal resistance is reduced, on airplanes, by means of wheels, and on seaplanes by the use of long, narrow pontoons.

Structural Materials Used: The choosing of material for aircraft construction is a study in itself. It is necessary to insure great strength and absolute reliability, without sacrificing lightness. The problem has been solved sufficiently to permit successful flying, but further study and research will doubtless result in marked improvement in the selection of material.

Wood is largely used in the construction of the framework. The chassis struts, skids, longerons, and engine-bearers are usually made of ash. This is a straight-grained, tough wood, but it is rather heavy. Spruce is extensively used for the main spars, wing spars, and struts. It is not so strong as ash, but it is considerably lighter and is quite dependable, and hence it is always used when it can be obtained in clear, sound, straight-grained lengths. Members are frequently built up of sections of ash and spruce, glued together. Ribs are usually made of white pine. Hickory is used for landing-gear struts, especially in the construction of heavy machines. Canadian elm is very tough and is sometimes used instead of ash for engine bearers and longerons. It is easily twisted and warped, however, and for that reason it is not wholly ideal. Basswood is used in the webs of ribs; and walnut, mahogany, and ash are used in propeller construction.

Aluminum is employed in constructing cowl supports, wind-screen frames, and control wheels. It is not particularly desirable in connection with seaplanes or flying boats, because it corrodes easily in the presence of water, and when it is used it must be constantly watched and frequently cleaned. Manganese bronze is quite tough, and on account of its resistance to corrosion it is used largely in airplanes operating from the water. It is also employed in making wood screws, and for bearings for rotating parts. Phosphor-bronze has properties resembling those of manganese bronze, and it is used for similar purposes.

Steel is used for sockets and control leads, and for wire and wire attachments;—in fact, for all kinds of airplane fittings. Some builders have employed steel for the entire framework, but such construction is heavy and difficult to repair. Steel tubes may be used as struts, and in some cases a stream-line cross-section is given to these tubes by the application of sheet metal or balsa wood, externally.

Duralumin—a special alloy of aluminum, copper, and magnesium,—has recently been used in aircraft construction for all the purposes for which wood has hitherto been employed, except for making propellers. This alloy is tough, strong, and light in weight, and is

Courtesy JL Aircraft Corporation.

AN ALL-METAL MONOPLANE.

likely to play an important part in the aircraft of the future. An incidental but important advantage associated with the use of duralumin in the construction of all-metal machines consists in the fact that when an accident occurs the metallic construction-members bend and crumple up, thereby absorbing part of the energy of motion of the machine and lessening the violence of the shock. Wood, under similar stress, splits and snaps, and the broken ends often cause serious injuries.

To obtain the best results with duralumin, the alloy must be heat-treated before working, by subjecting it to a temperature of from 350° to 380° C. and then quenching in oil or hot water. The metal is thereby rendered plastic, so that it can be easily forged, stamped, drawn, or rolled. After working, a final heat treatment is necessary in order to give the alloy its

maximum hardness and strength. For this purpose it
is heated to 500° C. or 520° C., quenched in oil or hot
water, and then allowed to stand for about a week,—at
the end of which time it will have become permanently
hard and durable. The heating is a delicate operation
and is usually performed by the aid of a salt bath
composed of equal parts of nitrate of sodium and nitrate
of potassium,—the mixed nitrates melting at a temperature that is materially lower than the melting
point of either one when used alone. Accurate ther-
mometers or pyrometers must be used in this work, and
the temperature of the bath must be closely watched
and carefully regulated; because if the metal is heated
above 550° C. it becomes permanently hard and brittle,
and its strength is also materially reduced. If heated

Courtesy Air Service, U. S. A.

AN ALL-METAL BIPLANE.

even a few degrees above 520° C. the alloy loses some of its desirable qualities. Equal care is required in the preliminary heat treatment used in preparing the metal for working, because if a temperature of 400° C. is reached during this annealing process, the metal becomes hard and difficult to work.

Linen and Egyptian cotton cloth are the principal fabrics used for covering the framework of the wings, and the fuselage or nacelle. Recently, thin sheets of corrugated duralumin have been used experimentally in place of fabric for wing and body coverings, and it is said that the metal has served this purpose admirably.

Control: There are three points of control in the standard type of airplane,—namely, (1) the ailerons located along the trailing edges of the wings and near their outer ends, (2) the elevators at the rear of the fuselage, and (3) the rudder (or rudders, when there are more than one) located in the rear of the body. These parts are operated from the pilot's seat by means of levers and wire cables.

The rudder is normally operated by means of a foot-bar, and is used for the same purpose as the rudder of a boat. A lever, called a control-stick or "joy-stick," is commonly used for controlling the ailerons and the elevators. In some cases, however, a wheel is employed instead of a control-stick. A fore-and-aft movement of the wheel or stick controls the elevators, and the ailerons are operated by rotating the wheel or by moving the stick sidewise. The elevators control the vertical movements of the airplane, while the ailerons, which control the movements that correspond to the "rolling" of a ship, are used principally in banking for turns. The ailerons are usually of the double-acting type, in which compensating wires are used. When the aileron on one side is deflected upward by means of the control lever, the opposite aileron is simultaneously lowered by means of the compensating wires. In other words, the construction is such that the elevation of one of the ailerons is necessarily attended by the depression of the opposite one.

The method of running the control cables, as well as the construction of them, is extremely important. They should always run through well lubricated leads, and every precaution should be taken against clogging or jamming. Sharp angles in the cables should be avoided by the use of chains or bell-cranks or other similar arrangements. All control cables should be flexible, and it is advisable to have them in duplicate, wherever possible. The main point is to insure absolute freedom and certainty of movement, because the safety of the airplane depends at all times upon the positive operation of its controls.

Certain special stabilizing devices, embodying the principle of the gyroscope, are used in some machines. To what extent they will be employed in commercial aviation (if at all) is thus far undetermined. Even the value of them is not yet universally conceded. Stabilizers of some forms can be set or fixed in position so that the airplanes in which they are installed will run in a predetermined course, subject only to wind-drifting, as long as the motor operates. Aviators have been known to start their stabilizers and then sit in their seats writing letters, paying no attention to their controls. This is an exceedingly unwise practice, because it is always dangerous to place too implicit a dependence upon automatic devices of any kind, especially when failure would be followed by disastrous consequences. The aviator should give his personal and immediate attention to the control of his machine at all times, and should never rely upon automatic apparatus. Such apparatus is useful in so far as it *assists* the aviator, but he must be watchful of his machine at every moment that he is in the air.

Plan and Performance: Airplanes of numerous makes and models are now produced commercially, and each of them has its own good points. As in selecting an automobile, the use to which the machine is to be put should be the chief factor in making a choice. There are certain points in airplane construction, however, that determine the usefulness of the craft for any purpose, and these should receive serious consideration.

In the first place, the airplane should be of a clean-cut design, and stream-lined, and the gaps between the control surfaces and the main sections should be small.

The stability required in a machine will depend largely upon the purpose for which the airplane is to be used. A certain amount of stability is essential to safety, but in low-powered aircraft excessive stability causes the machine to be subject to violent reactions in a gusty wind, and makes it hard to handle. When such planes have too much stability, they are extremely safe but very uncomfortable to ride in. Lateral stability is secured by a combination of rear fin area and wing dihedral; and longitudinal stability depends upon the location of the center of gravity relatively to the wings. The airplane will be unstable unless the horizontal tail surfaces are set at a negative angle with respect to the wings;—that is, unless the entering edges of the horizontal stabilizers are slightly lower than the trailing edges.

The ease with which an airplane can get off the ground depends primarily upon the design of the wings. A low load per horse-power will enable a plane to lift quickly, and ability to do this is also obtained if the load per square foot of wing surface is low. A propeller of small diameter is desirable in obtaining the maximum number of revolutions per minute from the motor while the plane is still on the ground, but is not desirable when the plane is in the air. A large propeller is then preferable, in order to obtain a maximum amount of power. Increased ease of "get-away" involves a corresponding sacrifice in lifting power or useful load.

Speed in the air, and climbing ability, are not paramount considerations in commercial planes, yet these characteristics are sometimes highly desirable when it becomes necessary to avoid trees, houses, smokestacks, and other obstacles near a flying field. If an airplane travels at its maximum speed a considerable part of the time, the engine is subjected to an enormous amount of wear and tear and it will not stand up long under the strain. A plane should have a flying speed compatible with the work it is engaged in, and still have some speed in reserve for use in an emergency. A high ceiling is desirable in a mountainous country, where flying at considerable altitudes is necessary in order to clear high peaks.

An airplane having deeply cambered wings is likely to possess great lifting power. Commercial planes should have wings that are moderately cambered, —a compromise between those of a speed plane and those of the extremely slow type. Low flying speed enables a plane to land with less shock, and also to come to rest soon after striking the ground. This is important in connection with planes that may have to land in restricted areas, but on the other hand too low a landing speed is undesirable on account of the reactions to which the plane is subjected in gusty winds. It is possible to reduce the landing speed by flattening out the dive, and landing with the speed slightly below the theoretical value.

The head resistance of the airplane and the drag of the tail skid reduce the distance a machine must taxi after landing. A large angle of incidence in the wings creates a large amount of head resistance. With the tail skid on the ground, a sixteen-degree angle of incidence between the wing surfaces and the line of flight is considered to be about the minimum value for general work. The center of gravity of the machine should be at least twelve inches back of the wheel axis of the landing gear, to resist the tendency of the machine to

Courtesy Curtiss Aeroplane and Motors Corporation.

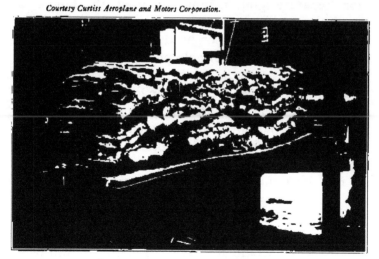

SAND-TESTING AN AIRPLANE WING.

"nose over." This position of the center of gravity also insures a reasonable amount of drag on the tail skid.

The factor of safety should be at least six in the construction-members of large, stable planes, but in smaller types that are subject to manoeuvers it should be increased to at least twelve or fourteen. The materials used in all the members of an airplane should be thoroughly tested, to be sure they are sufficiently strong.

In the power plant, high-compression engines are not considered as reliable as low-compression ones, especially when high compression is coupled with high piston speed. Most airplane engines are equipped with two sets of spark plugs, and it is still better to have two or more entirely separate ignition systems. Ignition trouble frequently causes engine failure, but if two or more separate ignition units are used, the failure of one of them will not force the pilot to stop flying. He can continue to operate on his duplicate or reserve unit until he can get to a suitable place to land and make repairs.

Multiple engines are desirable for the same reason; for if one of them fails, the pilot can usually continue flying under partial power long enough to reach a suitable landing place. Multi-motored airplanes require extra large and extra strong rudders, to compensate for the one-sided pull that is exerted if some of the engines are not working properly. Without reliable rudder control, the machine is likely to side-slip under these conditions.

It is important to have the gasoline feed-pipe arranged so that there will be no chance of accidental ignition of the fuel supply. The feed-pipe is frequently installed in such a way that it passes close to the ignition apparatus, where a leak would surely produce a fire. This is a serious error in design. Gasoline feed-pipes should be kept well away from all sparking devices, fuse blocks, and switches, and from all the highly heated parts of the motors. They should also be made of soft copper or other satisfactorily flexible tubing, so that the chance of breakage will be reduced to the lowest practicable point.

II. THE OPERATION OF AIRPLANES

INTRODUCTORY: The safe operation of an airplane depends largely upon two factors: (1) the efficiency and fitness of the pilot, and (2) the physical condition of the machine he is operating. To make sure that the airplane is in proper condition for flying, it is necessary for the pilot to inspect his machine thoroughly before leaving the ground. This inspection should be carried out in accordance with a definite plan or system, in order to be sure that none of the vital parts of the machine are overlooked.

Tuning-up: In making an inspection, it is good practice for the pilot to take a position on the right or left side of the fuselage, and then make a complete circuit of the machine, either to the right or to the left, carefully looking over and testing the various parts as he goes along. The fabric or wing coverings, cables, bracing wires, interplane struts, landing gear, tail skid, control surfaces, and control cables should receive special attention.

The fabric should be tight and free from rips and tears, and all small holes should be patched and well protected by dope or varnish. Loose fabric increases what is known as "skin friction," and retards speed in flying. Exposed fabric deteriorates with extreme rapidity, and airplanes should therefore be protected from the weather when not in use. Oil destroys the fabric varnish and for that reason any grease or oil that may be found upon the wings should be carefully and thoroughly removed.

The cables and bracing wires should receive a light coating of grease or oil to protect them from rust. Cables that show frayed or broken strands should not be repaired, but should be immediately *replaced by new cables*. The flying and landing wires should be under sufficient tension to keep the framework rigid, but it should be remembered that too much tension subjects the parts of the framework to needless strain, and thereby lowers the factor of safety that has been provided for the purpose of taking care of the unforeseen stresses to which the machine may be subjected in its regular operation. Examine all turnbuckles for loose or missing safety or locking wires. If these locking wires are absent or loose, the barrels of the turnbuckles may turn and release the tension on the bracing wires. See if any of the cotter pins or nuts have been lost, and replace those that are missing.

The interplane struts should be examined for splits, cracks, and warping. Warped struts are likely to break down under the pressure to which they are subject. The socket arrangements by which the struts are fastened are usually connected to the wing fixtures by pins or bolts, and these pins or bolts should be locked in position by means of split pins.

Wing skids are not always provided, but their

use is highly recommended. They are not needed in a perfect landing, but on rough ground, or in landing with one wing low, they are of great advantage in avoiding the possibility of a serious crash. They are specially valuable if the airplane has ailerons attached to the lower wing.

The landing gear struts should be well bedded in their sockets, because otherwise the shock of a rough landing will drive them in further, and in this way loosen the tension on the cross-bracing wires. This would throw the whole undercarriage out of alinement and might greatly weaken it. It would also prevent the machine from taxying straight. The shock-absorbers should be examined closely,—particularly if they are composed of rubber, because this material deteriorates even from mere exposure to sunlight, and is quickly destroyed by the action of lubricating oil or grease. Make sure, also, that the wheels are securely fastened to the chassis and that they are properly oiled. They should revolve freely, and the tires should be properly inflated.

The tail skid and post are frequently overlooked during an inspection, but they are important parts of the airplane and should receive proper consideration. They are probably broken and repaired more frequently than any other airplane parts. If the shock-absorber portion should become loose or permanently stretched, the shock of landing (which is relieved by the tail skid) would be received by the fuselage, and the whole framework would be likely to be severely strained.

The control surfaces should be examined to see that they operate easily and positively. The horns should be tight, and the hinges should be secure and in perfect condition, and the hinge-pins should be in place and properly locked. The cable attachments should also be carefully examined. The cables should be followed throughout their entire length, and any defective places should be noted and remedied. Lubricating oil should be applied freely wherever a lead passes through a pulley or sheave, and special care should be taken to see that the cable is not frayed at such points. Unnecessary slack in the cable leads will cause the control surfaces to act sluggishly, and will make it difficult to handle the machine effectively in an emergency.

The general examination of the airplane being completed, attention is next directed to the engine, and to the fuel and water supplies, the instruments, and the personal equipment. The supplies of fuel and water are normally taken care of by an attendant assigned to that duty, but it is well to check up his work, especially if a cross-country flight is contemplated. Personal inspection of the engine is not absolutely necessary, because defects in this part of the mechanism will probably betray themselves during the warming-up process. It is important, however, to look it over in a general way, and the aviator should at all events make sure that there are no leaks in the gasoline tank or the feed pipes, or see that any such leaks that exist are immediately repaired.

On entering the airplane, the safety belt should be adjusted promptly. Examine the instruments and make sure that the altimeter is set at zero. If the fuel is supplied to the engine under pressure, pump up the necessary pressure by means of the hand pump provided. Test the controls to see that they operate freely and effectively. The spark control, and the switches, gasoline shut-off, oil-pressure gage, air release, and other controls and appliances should be tried, and adjusted so that they work properly.

When ready to start the engine, see that the blocks are secure in front of the chassis wheels,—and be specially sure that the throttle is only partly open, because many serious accidents have occurred in consequence of turning over the engine while the throttle was open too wide. Orders given to the attendant should be clear and distinct, and they should be repeated by him before being carried out. It is advisable to pull the control stick well back, in order to prevent the tail from lifting and to avoid turning a somersault over the blocks.

The use of mechanical starters cannot be too strongly recommended, because they largely eliminate the accidents associated with starting. Various kinds of starters are available, ranging from small compressed-air devices to large separate cranking machines mounted on light motor-trucks. On commercial routes it has been suggested that the planes be equipped with starting motors, and that the batteries needed to operate these motors be mounted on small trucks that can be wheeled around to the different planes on the field as needed. This plan would provide electric starting apparatus and yet it does not require each plane to carry a storage battery.

Never run the speed of the motor up beyond 600 revolutions per minute, until the temperature of the engine is at least 60° Fahr. Undue speed when starting strains the motor and causes it to heat up too quickly, and quick heating is likely to warp the valves and prevent them from seating properly. After the motor has attained a temperature of 60° Fahr. it may be run faster, and the throttle may be slowly opened to make a full-speed test for vibration, and to see that the engine works properly. The throttle should be opened up gradually until it is wide open, and full speed should not be maintained for more than a few seconds. Then slow the engine down to about three-quarters of its maximum rate, and run it at that speed until you are thoroughly satisfied that everything is working properly, and you are ready to take the air.

If the engine misses fire while taxying to the "take-off" or if it does not work properly in any other respect, *do not "take off"*, but return to the starting point and test out the engine and make the necessary adjustments until it operates satisfactorily. An engine that does not work smoothly is likely to cause the machine to lose speed in taking off, and a sudden fall may result.

Standard Clothing and Equipment: The selection of clothing and other personal equipment is largely a matter of taste and circumstances. Provision should be made for cold weather, especially if an extended flight is contemplated. The regulation leather coat is desirable because it keeps out wind and moisture, and at the same time retains the heat of the body. The one-piece flying suit and sheep-lined flying moccasins are recommended.

Nearly all flyers wear goggles, because they afford comfort and protection for the eyes. If goggles are not worn, flying is likely to produce spasm in the eyes, and it sometimes brings on a special form of conjunctivitis. The lenses of the goggles should be constructed so that the eyes will not be damaged by splinters of glass in case of an accident.   One well-known method of achieving this end is to make each lens of two thin pieces of glass, with a film of transparent celluloid cemented between them.   In case of fracture the broken pieces then adhere to the central layer of celluloid, and the danger to the eyes is thereby materially reduced.   The nose-piece of the goggles should be made of some soft, non-metallic substance, that will not injure the nose.   Safety helmets are valuable for protecting the head from injury in a crash.   They should be light in weight and should fit properly, so they will not be dislodged easily.

III. AIRPLANE ACCIDENTS

GENERAL CAUSES OF ACCIDENTS: An airplane accident is hardly ever due to a single cause. Usually several factors are involved, among which may be mentioned structural defects, engine trouble, error in judgment, physical illness, fatigue, and "loss of head." If only one of these factors influenced the situation, an airplane could usually be landed safely; but if (for example) the engine stops, the pilot is likely to "lose his head" and do something he should not do, or, if he is fatigued, his condition will probably affect the soundness of his judgment and the result may be quite serious.

A survey of the accidents that occurred during a period of six months at one flying field, shows that in 9,000 flights there were fifty-eight crashes, in which sixteen men were seriously injured. Four of these accidents were considered unavoidable, and only one was caused by the failure of the airplane. The remainder were due to the ineptitude of the pilots,— forty-two being caused by lack of judgment, seven by loss of head, and four by fatigue.

Courtesy Air Service, U. S. A.

AIRPLANE AFTER TURNING TURTLE.

(Poor judgment is responsible, nine times out of ten, for such accidents.)

In their relative proportions, these figures are fairly representative of the experience on most flying fields, both in Europe and in the United States. They serve admirably to show the importance of having the very best type of pilot obtainable;—one who is physically, morally, and mentally fit, and competent in every way.

Errors of the Pilot: An error in judgment is perhaps the most common cause of airplane accidents. A pilot frequently misjudges his distance from the ground when landing and flattens out too soon or too late, and an accident may easily occur before the mistake is realized. In the air, he may bank too much or too little or he may climb on a turn. Accidents from attempting to make turns too near the ground, where a side slip means disaster, are notably frequent.

A pilot often subconsciously senses a danger to which he is subjected, but under the sudden strain of the emergency he may be unable to think, decide, and act quickly. This momentary lapse of co-ordination between reason and action is called "loss of head." In flying, a fraction of a second often counts greatly, and may be all that stands between safety and danger; and in an emergency there is seldom time to correct an error.

Loss of head is closely associated with fatigue and with fear. When fatigued, a pilot is unable to think and act quickly, because his brain no longer responds, with its normal promptness, to the demands that are made upon it. The pilot works in a sort of stupor, and takes but little conscious part in controlling his plane. If a crash occurs from this cause, and the pilot escapes, he usually has no distinct recollection of what happened during the flight;—his memory appears to have been temporarily stunned.

Fear is closely associated with fatigue and loss of head, but it does not appear to produce many accidents. There is little time to think of danger during flight, and consequently, even though there may be a sense of fear lurking somewhere in the back of a pilot's head, it rarely asserts itself in such a way as to affect his management of the machine. When it does, however, the effects are similar to those produced by loss of head and fatigue.

Courtesy Air Service, U. S. A.

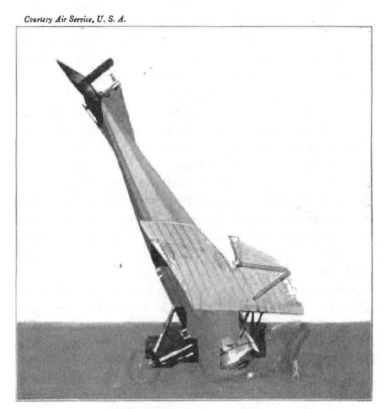

A RESULT OF DEFICIENT VISION.

Failure of the Machine: In the early days of flying, accidents were frequently caused by the failure of some vital part of the plane, but this difficulty has now been largely overcome. There is, however, a great need of inspection before attempting to take a machine off the ground. During flight the plane is subjected to severe strain and vibration and its parts are likely to become worn or loosened. A pilot should always see that his airplane is mechanically safe before he attempts to use it, and the subject of inspection is discussed in some detail in another section of this book. (See page 36.)

Fire: It is not uncommon for a plane to catch fire in the air, and investigations of several accidents of this nature show that defective gasoline feed systems have been the primary cause. It has developed in some cases that inadequate drainage of the fuselage has allowed an accumulation of gasoline immediately under the engine, and the fumes from this exposed fuel catch fire, either by a back-fire, or from an insufficiently protected exhaust manifold, or from a spark from one of the magnetos.

There is always more or less free gasoline around an aeronautical engine,—coming from a flooded carburetor, a leaking feed-pipe, or some other source. Ample drainage facilities, in the form of fair-sized drainage holes in the fuselage bottom, should be provided so as to allow the oil and gasoline to drain out freely when the machine is in any possible position. The entire gasoline supply system should be inspected for leaks before every flight, and any leaks that may be found should be repaired before the flight is attempted. It has been demonstrated that there is less likelihood of leakage occurring in the gasoline feed system if the supply tanks are connected to the other parts of the system by means of flexible tubes of copper or other material that will withstand vibration.

To remove the fumes of gasoline, the engine compartment should be adequately ventilated. In spite of good ventilation, however, there is always a possibility of inflammable vapors remaining near the engine, and to prevent ignition of these fumes by back-fires it is necessary to carry the open end of the carburetor air-intake outside of the engine compartments. With this arrangement flames from a back-fire are blown out into the open air instead of into a space that may be filled with inflammable fumes.

As a protection against the accidental ignition of inflammable fumes and vapors by sparks from the magnetos, it appears to be possible and practicable to inclose the magnetos on aeronautical engines by means of gauze covers similar to the kind used in safety lamps and on explosion-proof motors and dynamos. Protection of this kind might not prevent fire in case the magneto became drenched with liquid gasoline, but it would almost certainly prevent the accidental ignition of inflammable vapors in the engine compartment, so long as the gauze remained whole and sound.

To prevent a fire from attaining serious proportions, a pressure-actuated sprinkler system has been found efficient in many cases. Such systems operate on a principle similar to that used in connection with the sprinkler systems found in large buildings, except that the fire-extinguishing medium in the airplane is not water, but pyrene, fire-foam, or some other material effective in quenching oil-fires, and that the system is

From Underwood & Underwood, N. Y.

A Typical Airplane Fire.

actuated by air pressure. A tank is installed in the airplane, and small pipes or tubes are run from it to various parts of the engine compartment. A valve, actuated by the release of a fuse of soft alloy, is provided, and when a fire breaks out the fuse melts and instantly the entire engine compartment is flooded with an effective fire-extinguishing spray. The main objection to the pressure fire-extinguishing system is that it makes considerable additional weight for the airplane to carry, but this objection is offset by the protection that the system affords.

The danger from fire is not confined to the period of actual flight, for it must be remembered that when a "crash" occurs the aviator may be pinned down by the wreckage or rendered helpless in some other way, and he is then in great danger if the wrecked machine takes fire.

Airplane fires that occur in consequence of crashes are caused largely by the bursting or puncturing of the gasoline tanks, or the rupture of the tubing, from the violence of the impact. In most airplanes it is necessary to carry the gasoline under pressure, in order that the fuel may reach the engine when the machine is "nosed up" at a considerable angle. When the gasoline tank is perforated at any point below the level of the liquid surface, the air pressure forces the fuel out in a fine spray, and the pilot, passengers, and machine are likely to become drenched with it. If, as frequently happens, this fuel becomes ignited by a spark from any source, the results are usually extremely serious.

Experiments have been carried on with the intent of producing a tank that will not burst or puncture in a crash, or which will not distribute the gasoline over the machine in case of an accident. These attempts have been fairly successful, and safety tanks of various kinds are now available. In the main, safety gasoline tanks consist in a metal shell of medium thickness, covered with fabric and vulcanized rubber of varying degrees of elasticity. The whole is further covered with galvanized wire netting. The idea is to provide a flexible form of construction that will withstand severe shock. The tubing used in connection with these safety tanks should be of soft copper or other flexible material. It is altogether probable that many of the fatal airplane accidents, in which the serious features have been due to the outbreak of fire after the crash, could have been prevented or rendered far less serious if safety tanks had been installed on the machines.

To prevent the spreading of fire in an airplane, the dope used on the fabric should be as nearly fire-proof as possible, and to insure greater safety, the cloth also should be fireproofed before the dope is applied to it. Fire-resistive dopes have been produced in various ways, the most common method being by the addition of certain fire-retarding substances to the ordinary acetate dope. Many of the dopes prepared in this way are objectionable on account of the fact that they are much heavier than ordinary dope and consequently their use materially increases the weight of the doped surface. Recent developments in some of these fireproofing methods, however, have reduced the added weight to a negligible quantity.

The use of oils and varnishes in finishing the woodwork in airplanes increases the fire hazard in these parts considerably, and the use of fire-resistive paint on all interior wooden parts, and especially in the engine compartment, is highly desirable. Fireproofing materials may be a little more expensive than the materials ordinarily used, but the added protection that they give appears to be well worth the difference in cost.

Superchargers and Variable-pitch Propellers: For use in flying at high altitudes, where the air pressure is considerably below normal, superchargers and variable-pitch propellers have been developed. Superchargers compress the air that is used by the engine, and deliver it to the cylinders under a pressure that is approximately the same as that prevailing at sea level. Propellers with a variable pitch can be so changed that their effect on the rarefied air at high altitudes will be practically the same as that of normal propellers on the air at sea level. Variable-pitch propellers might also be of considerable advantage in landing, because by reversing the pitch the head resistance of the airplane could be greatly increased, and the machine could be brought to a stop within a short distance after the wheels touch the ground. The practicability of the variable-pitch propeller is questioned by some authorities, however, on the ground that any mechanism that would vary the pitch of a propeller would in all probability tend to reduce the solidity of the propeller as a whole. This is a matter worthy of serious consideration.

Instruments: Although well trained and experienced pilots and mechanically perfect machines are the first requisites of safety in flying, various other factors are also of great importance in this connection. For example, dependable instruments are needed, to keep the pilot informed with respect to the speed, altitude, attitude, and direction of motion of the airplane.

The speed of the airplane is read from an *air-speed indicator*. This instrument indicates the speed with which the craft is moving, relatively to the air through which it passes. In still air and in low altitudes the air-speed meter also indicates the ground speed of the craft with fair accuracy, but if the wind is blowing the ground speed must be calculated from the reading of the instrument and the velocity and direction of the wind.

The direction in which the nose of an airplane is pointed is indicated by a *compass*. This instrument also enables a pilot or passenger to locate objects on the ground by bearings, when the position of the plane is known; and it is likewise employed, to some extent, for determining the position of the airplane itself, by taking cross-bearings upon known objects. The compass is one of the most essential of all airplane instruments, and one that is most likely to give false information unless particular attention is given to its installation, and unless the readings are taken while the machine is flying level and on a straight course.

The *altimeter* is used for determining the height of an aircraft above the surface of the earth. This instrument is usually of the aneroid barometer type, and may be either indicating or recording in its operation. A bubble *statoscope* is also desirable, to indicate short, rapid changes in altitude, too small to be shown clearly on the altimeter. Its use assists a pilot in holding his machine at a constant level.

The purpose of the *inclinometer* is to show at what angle the airplane is flying,—indicating the lateral as well as the longitudinal angle that the plane makes with the horizontal. In order to insure the effective operation of an inclinometer, the instrument must be stabilized by a gyrostat or other equivalent means, and all readings must be made when flying in a straight line at a uniform speed.

In planes equipped with radio apparatus, the *radio direction finder* is rapidly coming into use. In operating this device, closed-coil aerials, fastened in the wings of the machine, are used. A closed flat coil possesses strong directional characteristics, because when the edge of such a coil is pointed directly toward the incoming electrical waves, the signals received are of maximum strength; but as the coil is turned to one side or the other, the signals rapidly become weaker and less distinct. With a radio direction finder it is possible to guide an airplane with considerable accuracy toward any selected radio transmitting station, even though the station is entirely invisible, on account of distance or bad atmospheric conditions.

Safety Straps: It should hardly be necessary to emphasize the importance of using safety straps in the seats of airplanes, but experienced aviators not infrequently fly with their safety straps undone. Such practice is foolhardy, yet it would be easy to mention some distinguished aviators who have been killed by carelessness in this respect. It is absolutely essential that the pilot and passengers in airplanes be strapped to their seats, to prevent falls when the machine turns on its side or on its back. The straps that are used should be broad and exceedingly strong, and be securely fastened to the framework of the machine. The device employed for fastening the belt around the person using it should be constructed so that it can be released quickly and with one hand; and it is recommended that this release be effected by means of a small hand lever, located where it will be easily accessible under all circumstances, but where it cannot be operated accidentally.

Emergency Stations: Since only a small number of airplane accidents occur outside of landing fields, the airdromes are the scenes of the greatest catastrophes. A surgeon should be employed at every permanent landing field, to furnish assistance in case of accidents. Every airdrome should also maintain an emergency station and an ambulance, and a number of men trained in first-aid work should be available to assist in treating injured persons.

The emergency station should be well supplied with materials necessary for treating injuries of all kinds. The ambulance should carry a supply of stretchers, bandages, splints, surgeons' plaster, field dressings, slings, morphine, hypodermic syringes, anesthetics, and anesthetic face-masks, as well as scissors, knives, and other instruments that may be needed by a surgeon in field work. Wirecutters (suitable for cutting airplane wires), saws, hammers, crowbars, and other tools that may be needed for clearing away wreckage should also be carried on the ambulance. Fire extinguishers are likewise essential.

In the event of a crash or a bad accident of some other kind, the surgeon and a staff of first-aid men and mechanics should proceed with the ambulance to the scene of the accident, at the earliest possible moment. The persons involved in the crash should be removed from the wreckage at once, and placed on stretchers if they are injured. The surgeon should make a rapid examination of the injured persons and direct such first-aid treatment as he deems advisable. Only such treatment should be given on the field as is necessary to relieve pain and to make the removal of the patient safe.

In releasing an injured person from the wreckage, cut away the debris that is holding him down, instead of trying to drag him out. Pulling persons from the wreckage may convert simple fractures into compound ones, and add materially to the seriousness of the injury. In case of fire, use the fire extinguishers on the parts near the injured or imprisoned persons, and *be careful to direct the streams in such a way that the injured persons will not be suffocated by the vapors.*

After the injured and other persons have been removed from the wreck, a corps of men should be assigned to take the damaged plane from the field. This should be done as soon as possible, because any obstruction remaining on the field may seriously interfere with the safe operation of other airplanes using the airdrome.

IV. PILOTS

**THE IMPORTANCE OF LEGAL REGULA-
TION:** In the earlier days of aeronautics, little
consideration was given to the qualifications that a man
should possess, to become a flyer. Anyone who was sufficiently daring and self-possessed was considered fit for
the work, and no other special characteristics were
thought to be necessary. The result was, that many
accidents were unexplained and there was an enormous
avoidable waste, both of men and of machines.

Under present conditions, it is not a difficult matter
to obtain an airplane-pilot's license from the International Aeronautic Federation,—an organization founded
in 1905 for the purpose of regulating aeronautics, but
confining its activities to the control of aeronautic
sports. There is no other civilian body in the United
States that issues or requires licenses at the present time,
save in a few states and municipalities where laws or
ordinances, suggestive of those established in connection with automobile traffic, have been enacted for
the local regulation of aeronautics. With these excep-
tions, and with the important additional exception of

the United States Air Service, the operation of aircraft
and the licensing of pilots are nowhere officially controlled or provided for.

Relief from this highly unsatisfactory condition
of affairs is promised for the near future, however.
The United States has declared its adherence to the
International Convention Relative to Air Navigation,
and as a sequence to this action Congress will doubtless
take appropriate action for establishing an Air Navigation Commission, and drafting rules and regulations
affecting aerial navigation in general. When this has
been accomplished, air pilots will probably be licensed
by the Federal Government, in accordance with some
definite plan yet to be determined.

Pending the establishment of a national bureau
for carrying on this work, we offer, below, some con-
structive suggestions which may be useful to local
authorities who are desirous of taking immediate
action with regard to air navigation and the licensing
of aerial pilots.

Physical and Mental Qualifications of Pilots:
Flying does not require a super-man, and in fact a
super-man is undesirable. A flier must be normal in
every way and any variation from this condition re-
duces his ability to manage a flying machine. Many
authorities assert that if the machine is properly
controlled, flying is not much more hazardous than
riding in an automobile; but even if this were true, we
must surely admit that the act of providing this control imposes unique demands upon the pilot. He is
the heart and brain of the airplane and it has been
said that no other occupation subjects a man to
strains as varied and intense as those that he sustains

Courtesy Air Service, U. S. A.

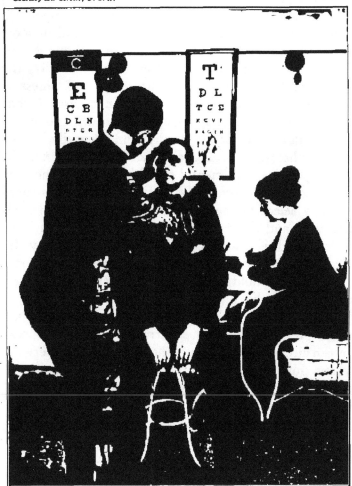

TESTING A CANDIDATE'S EYES.

while operating a heavier-than-air flying machine. Moreover, he is working in an unnatural environment, and is almost wholly unaware of its effect upon his nervous system. The machine itself may fail in some part and still be brought to the earth without serious injury; but if the pilot relaxes even momentarily, the whole machine is without a director,—and unless it possesses a degree of stability far in excess of that usually provided, it may easily crash to the ground.

A pilot must be physically and mentally fit for his work before he is taught to fly, and he must keep in proper physical and mental condition all the time that he is engaged in aeronautical work. The importance of properly selecting the men who are to engage in such activities, and the desirability of keeping these men in perfect condition, have been fully demonstrated by the Medical Department of the United States Air Service.

A pilot should be at least nineteen years of age, and he must be physically perfect in every way,— showing no abnormality, congenital or otherwise, that might prevent him from effectively and safely operating an aircraft. His heart, lungs, kidneys, and nervous system must be sound and healthy and capable of withstanding the effects of prolonged flight and of rapid changes of altitude. Moreover, his family history should show no inherent ailments or diseases of a nervous type, which might develop quickly in his own case and cause a temporary or permanent mental collapse.

His various special senses should be normal in every way. His eyes should show normal stereoscopic and color perception, and his general field of vision should be good. Persons whose eyes show more than two dioptrics of hypermetropia (far-sightedness) or myopia (near-sightedness) should be rejected. The middle ear should be healthy, and the vestibular apparatus should be intact and neither more nor less sensitive than the normal. The nose should show free air passages on both sides, and there should be no evidence of any serious acute or chronic affection of the upper respiratory tract.

Training: If the candidate is found, by examination, to conform to these requirements, he is ready for his aeronautical training. This should begin with technical instruction, to teach him the principles involved in flying and to make him entirely familiar with the construction and operation of aeronautical engines and airplanes. This work should be thorough, and it should be done under the personal guidance and supervision of an experienced mechanician.

When the technical training of the candidate is satisfactorily completed and he thoroughly understands the operation of airplanes and aeronautical engines, he may be taught how to fly. In connection with flying instruction, an ingenious special training apparatus known as an "orientator" has been found to be extremely useful. It consists in an airplane cockpit suspended within concentric rings or gimbals, in such a way that a person in the pilot's seat can execute any manoeuver that can be accomplished with an airplane, except a motion of translation. He can spin around in any way whatsoever, but cannot move in a straight line either forward, backward, sidewise, or up or down. This machine can be used by the aviator, with entire safety, in acquiring a tolerance for vertigo, and in learning to adapt himself to the rapidly changing conditions that are experienced while flying.

Courtesy Ruggles Orientator Corporation.

AN ORIENTATOR IN ACTION.

The use of the orientator in flying schools promises to materially shorten the time of flying instruction, and to save many lives and many machines.

The actual flying instruction, in the free air, should be given in a dual-control machine, and under the direction of an expert flyer. At first the candidate should do but little of the actual operating, and he should never be allowed to manage the machine throughout the entire flight, until he has thoroughly proved his efficiency in controlling it. Before being allowed to "solo" (that is, to fly alone), the student-pilot should make quite a number of flights in which he does all of the actual manipulating of the machine, including the preliminary inspection, the take-off, spirals, turns, spins, glides, dives, and landing, accompanied by his instructor but receiving no assistance from him. When the novice has demonstrated in this way that he is competent to fly, he should be allowed to take a machine up-alone; but his flying should be confined to the vicinity of the airdrome until he has completed at least eighty-five hours of solo work.

Examination and Licensing: When the student-pilot has eighty-five hours of solo flying to his credit, and has made not less than eighty safe landings, he should be ready to qualify for a commercial pilot's license. He then presents himself to be re-examined physically by a competent medical board, and he should also undergo practical tests at the hands of an examining board composed of expert flyers. If these boards both find the candidate physically qualified, and competent to manage an airplane, a license, bearing the date of issue and valid for six months (except in case of sickness or accident), may be issued to him.

He should be re-examined every six months thereafter, however, and if he is still found to be physically fitted to fly, the date of the re-examination should be recorded on his certificate. Re-examination should likewise be made after every illness or accident that the aviator may experience, and in all such cases he should be pronounced physically and mentally qualified to fly, before being allowed to resume his aerial duties. No pilot's license should be valid for more than six months from the date of the last physical examination.

Care of the Pilot's Health: In the larger permanent airdromes, where a number of pilots are on duty or in training, a certified physician or flight-surgeon should be employed to supervise the recreation and physical training of the aviators. He should study the habits, temperament, and general fitness of each individual flyer, and act as a medical advisor to whom the men may turn for counsel in time of need. An aviator may be suffering from some temporary mental disturbance, for example, or he may be slightly out of condition physically, and to fly under such circumstances might spell disaster. In cases of this kind the counsel of the flight-surgeon should be sought and his advice followed.

The aviator, if he is to maintain his highest efficiency, must be careful as to what he eats and when he eats it. It is advisable to provide a special eating place for pilots, and to have the food that is served to them prepared under the direction of some person who thoroughly understands dietetics and food values. College athletes have "training. tables" provided for them, and an aviator surely has far greater reason than they, to keep himself in perfect condition.

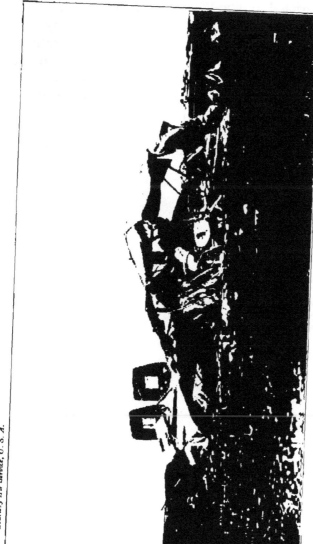

Courtesy Air Service, U. S. A.

THE CRASH IN WHICH RESNATI MET HIS DEATH AT MINEOLA.

(This great flyer was sick, and was advised not to fly on the day this accident occurred.)

Rational and well-considered exercise is essential to the maintenance of physical and mental alertness, and for this reason special flying calisthenics, particularly adapted to the needs of aviators, have been devised. These exercises should be executed at least once every day, by all pilots, and preferably under the direction of the flight-surgeon,—not primarily for muscular development, but to promote rapid and accurate coordination in the pilots' mental and physical activities.

V. THE MAINTENANCE AND REPAIR OF AIRPLANES

THE REPAIR SHOP: Emergency repairs are usually made by the airplane pilot or his mechanic, but the constant strain and wear caused by continued operation also cause rapid general deterioration, and this makes it necessary to overhaul every airplane frequently and thoroughly. The engine should be fully inspected in all its parts after every 50 hours of operation, and it should be completely overhauled after it has operated from 100 to 150 hours. For work of this kind the airplane is sent to the repair shop.

Repair shops do not, in general, include facilities for making large castings or intricate parts, but small aluminum castings, and parts that are ordinarily made in a machine shop or blacksmith shop, can usually be turned out in the airplane repair shop.

The parts kept in stock should include bolts, nuts, cylinder studs, piston rings, rocker arms, hinge pins, metal sockets, control-wire guides, longeron clips, shock-absorber guards, and in fact all of the small fittings used in airplane construction.

The woodworking section of the repair shop should

be equipped to make any of the wooden members found in a plane, including longerons, interplane struts, rudder posts, tail skids, wing skids, spars, compression and former ribs, and aileron beams.

In repairing airplanes, a systematic course should be followed. When a plane comes into the shop for overhauling, the engine should be removed and sent to the machine shop. The rest of the airplane should be thoroughly inspected and tested, and parts that are found to be weak or otherwise defective should be removed and replaced by new material. Worn-out or deteriorated fabric should be torn off and replaced by new. The landing gear should be carefully examined and rebuilt if necessary. When finished, every part should be tested and inspected with extreme care.

When the repairs are complete, the wings are

Courtesy Col. T. H. Bane and "Mechanical Engineering."

TESTING A LANDING GEAR.

Courtesy Col. T. H. Bane and "Mechanical Engineering."

SAND-TESTING A FUSELAGE.

alined and the plane is assembled. The airplane is then given the correct stagger, incidence, and dihedral, and is thoroughly air-tested.

In the machine shop, the engine is taken down and the various parts are thoroughly washed and cleaned. Every part is then inspected and repaired or replaced, as is necessary. When this has been done the engine is rebuilt and thoroughly tested on the testing block.

Repair Shop Hazards: The making of extensive repairs on airplanes involves practically the same hazards as those encountered in airplane manufacturing; but owing to the fact that repair shops are usually smaller than the factories, the hazards of one department are likely to be associated more or less intimately with those of another, because the workrooms are close together, and in some instances the various operations may be performed in the same room. The hazards that exist in shops under normal conditions were increased, during the war, by the necessity of realizing a high speed of production, by the unusual amount of night work that had to be done, and by the impossibility of exercising due discrimination in the employment of labor. In fact, the exigency of the times led manufacturers to close their eyes to many unsafe practices and conditions that would not be tolerated under normal or usual conditions. With the return of peace the continuance of hazards of this kind became unjustifiable, and there is no longer any good and sufficient reason why our workshops should not be made reasonably safe in all respects.

General Fire Prevention: There is considerable danger from fire in an airplane repair shop, and the entire building should therefore be equipped with a powerful and efficient sprinkler system, and with adequate standpipes and hose. Hand extinguishers should also be provided at numerous points about the workrooms, where they will be handy and available at all times. It is important to see that the water supply is fully adequate to meet any emergency that may arise, and that a good water pressure will be constantly available.

To prevent the spread of fire, it is advisable to subdivide each building into working areas as small as the nature of the work will allow. This may be accomplished by building fire-walls where space permits; and in other places, where the floor area cannot be subdivided, fire screens may be placed in the roof trusses,—extending from the base of each truss to the roof planking. These screens tend to prevent the spread of flames along the roof, and they also reduce the horizontal drafts; and in both these ways they materially retard the progress of a fire.

Woodworking: The framework of an airplane is usually constructed of small wooden parts, and the repairing of this skeleton involves the hazards usually associated with woodworking processes. The amount of waste is abnormally large, however, because only perfect material can be used, and considerable quantities of undesirable stock are therefore rejected.

It is extremely important to remove all waste material from the workrooms as fast as it is produced, and before it can accumulate in any quantity. Blower systems and dust collectors should also be installed to remove the sawdust and fine shavings, and dust-

Courtesy "U. S. Air Service."

A WELL-REGULATED WORKSHOP.

collecting hoods should be placed on all machines that produce dust.

The wood-storage area should be well separated from the main buildings, and at a safe distance from railway tracks upon which steam locomotives operate. The storage areas should be provided with ample hydrant facilities, and equipped with suitable hose.

The kilns that are used for drying the wood should be separate from the main building, and the practice of drying lumber in lofts over the boiler-rooms should be abolished. Kilns should preferably have brick walls and metal roofs, and they should be fitted with stout racks of steel or iron, for the wood to rest upon while drying. Caul boxes should be of metal, or of wood with a metal lining.

The drying material should never be allowed to rest upon the steam-pipes; and the practice of using an extended smoke-pipe for heating the cauls is dangerous and should not be allowed. In kilns and caul boxes the wood should always be kept at least twelve inches away from the surfaces from which the heat is obtained.

The airplane framework is held together largely by gluing, and this frequently involves a fire hazard in the use of glue heaters. Flame heaters should be eliminated if possible, and steam or electric pots should be used for heating the glue. The use of glue heaters can be avoided by using casein glue, and casein glue is rapidly superseding fish glue in the better class of airplane work.

In repairing the fabric covering of the wings and body of an airplane, a considerable amount of light, combustible material must be handled, in doing the necessary cutting, sewing, and fitting; and this means that there is a considerable fire hazard in this department. Work involving the use of fabric should therefore be carried on in an isolated building, or at all events in rooms separated from the woodworking rooms by fire walls.

Doping: For the purpose of making the fabric coverings taut and waterproof, they are covered with a special sort of varnish, after they have been fitted in place on the airplane framework. This varnish (or "dope" as it is called in the trade) varies in nature and may be divided into three kinds: (1) cellulose acetate dissolved in a solvent containing more or less tetrachlorethane; (2) cellulose acetate dissolved in a mixture containing no tetrachlorethane but consisting mainly of methyl acetate, methyl-ethyl ketone, acetone, diacetone, alcohol, and benzol; and (3) cellulose nitrate dissolved in a mixture of butyl acetate, ethyl acetate, alcohol, and benzol, or in other mixtures containing varying amounts of acetone, amyl acetate, alcohol, methanol, and benzol. Tetrachlorethane is not much used as a dope solvent at the present time.

When dry, the nitrate dope has properties somewhat similar to those of an ordinary moving-picture film. It burns with great rapidity, and if too highly nitrated it may also be explosive. The acetate dopes are far less inflammable when dry, but on account of the inflammable nature of most of the solvents that are used, they burn fiercely when in the dissolved state.

The greatest danger from the use of dope lies, however, not in the fire hazard associated with it, but in the poisonous nature of certain of the solvents that are employed. Tetrachlorethane, for example, is extremely dangerous, and its fumes seriously affect the liver and kidneys and the muscles of the heart. In fact, tetrachlorethane is one of the most poisonous of the chlorine derivatives of the hydrocarbons, and permanent destructive changes in the liver, through fatty degeneration, are more marked in connection with tetrachlorethane inhalation than with any other substance except phosphorus.

Benzol probably ranks next to tetrachlorethane in its harmfulness,—severe chronic poisoning from this substance invariably producing extensive destruction of the white corpuscles of the blood, and not infrequently giving rise to fatty degeneration of the liver, kidneys, and other internal organs. Benzol poisoning is characterized by a loss of weight and appetite, a quick, feeble pulse, a bluish appearance of the skin, digestive disorders, general weakness, and a tendency to fatigue after slight exertion.

Methanol (wood alcohol) is poisonous, and it causes dilation of the pupils of the eyes, blurs the sight, and produces abdominal cramps, nausea, chills, and drowsiness. Instances of total blindness from methanol poisoning are numerous, and fatal cases due to its absorption are not uncommon.

Chemically pure acetone fumes are said to be practically harmless when inhaled in moderation; but the fumes arising from impure commercial acetone cause headache and a burning sensation in the eyes.

Amyl acetate has a slight toxic effect, producing a smarting of the eyes, dry throat, sensations of tightness in the chest, and a tendency to cough. It also gives rise to serious nervous and circulatory symptoms including intense pain in the head. On account of the high cost of this substance it has been largely superseded by butyl acetate and ethyl acetate.

Amyl alcohol is said to be four times as poisonous as methanol, and its toxic properties have been estimated to be five times as great as those of ethyl alcohol (*i. e.* grain alcohol). It acts on the central nervous system and also produces a decided drop in blood pressure.

On account of the exceedingly poisonous nature of tetrachlorethane, dope containing this substance should never be used. It would be advisable to discontinue nitrate dopes also, on account of the fire hazard associated with them, though it is not likely that a radical change of this sort can be brought about in the near future. Doping with acetate dope, without the use of tetrachlorethane, may be carried on without any extreme hazard if a proper and effective ventilating system is used in the rooms where the work is done.

Such a system should be of the exhaust type, with powerful suction fans to draw the fumes out of the room. The suction outlets should be located *in or near the floor*, and it is best to make the floor of slat-like construction, and to draw out the fumes through the spaces between the slats. The vitiated air should be discharged into the open. Inlets for fresh air, having an aggregate area equal to at least three times the area of the discharge openings, should be provided at points remote from the outlets, and at least ten feet above the floor level. Men and women employed in the doping room should be instructed to conduct their work in such a way that the fumes and vapors will be drawn away from them. For example, in applying the dope the workmen should begin at the part of the machine that is nearest the exhaust outlet (if that is so situated that there is a horizontal current of air in the room), and work towards the incoming fresh air. It is just as easy to do this as it is to work in the opposite direction, and by attending to this simple precaution the fume-exposure can be materially reduced. The safest places in the room are the regions near the fresh-air inlets, and the workers should not be allowed to linger needlessly where the fumes of the dope solvent are strong.

Waste material, brushes, containers, and wipe-rags should be handled with care and kept at a safe distance from every heat source, to reduce the chance of their catching fire. The electrical equipment and wiring should be so arranged and protected as to prevent the possibility of fire, and no open switches nor fuses should be installed in the workroom. Vapor-proof globes should be used around the incandescent lamps, and it is safest to locate these lamps out-of-doors, so that the light from them will enter the workroom through the windows. It is best to avoid the use of pulleys and belts in doping rooms, but if they must be used the machinery should be effectively grounded to prevent the generation of static electric sparks. It is safest to ground the machinery in any event, and if electric motors are present they should be of the special vapor-proof type.

Machine Shop: It will not be necessary, in this place, to discuss in detail the general hazards involved in ordinary machine shops, because we have already considered them in another special book entitled *"Safety in the Machine Shop,"* copies of which may be had upon application to THE TRAVELERS

INSURANCE COMPANY. We shall therefore confine our attention, here, chiefly to certain special hazards that are encountered in airplane repair shops.

In repairing airplane motors, gasoline is used in considerable quantities, for cleaning the engines and engine parts. Dip tanks and spraying systems are employed, and there is often a copious evolution of dense gasoline fumes, which not only produce a bad fire hazard, but also gravely endanger the health of the workmen.

The cleaning should be done in special, isolated rooms, ventilated by a forced exhaust system, and but few occupants should be allowed in any one room. Special precautions should be taken in the installation and operation of electrical apparatus, as described above, and reliable means for extinguishing oil and gasoline fires should be provided. The use of respirators by the workmen is highly recommended.

Courtesy Col. T. H. Bane and "Mechanical Engineering."

TESTING AN ENGINE.
(This engine is connected to an electric dynamometer)

Motor Testing: When a motor is to be tested, it is placed on a testing block or airplane chassis and a propeller is attached. Guards should be installed in front of the propellers, because the rapidly revolving blades are extremely dangerous and many cases are on record in which workmen have been seriously injured by being struck by them.

As a number of machines may be undergoing test simultaneously, oil and gasoline are often present in large quantities. Back-fires are numerous and they are likely to start a conflagration. Whenever possible, testing areas should be located in open fields, at considerable distances from the nearest buildings.

When inclosures are provided for testing purposes, each stand should be separated from those adjoining it on either side by brick or concrete walls, and each stand should be covered by a roof of metal or other fireproof material,—the ends being left open. When necessarily located inside a building, the testing areas should be as small as practicable, and separated from the adjoining portions of the building by safe and substantial fire walls. Hand-operated fire extinguishers, specially adapted for use in connection with oil and gasoline fires, should be maintained in each compartment. Constant care should be exercised to keep the area as free from combustible material as possible, and all supplies of oil and gasoline should be removed as far from the testing areas as feasible, and should be stored and handled in a safe and approved manner.

In some instances, aeronautical engines are tested out by connecting them to an electric dynamometer. Such a method is used to determine the horsepower that a motor will actually develop while in operation.

VI. LANDING FIELDS, AIRWAYS, AND AERIAL LAWS

LANDING FIELDS IN GENERAL: Landing fields
may be divided into two main classes, (1) airdromes and (2) emergency fields. Permanent airdromes are large fields provided with hangars,
repair shops, gasoline stations, and accessories of
various sorts. Emergency fields are smaller than airdromes, as a rule, and are used only in case of forced
landings. They should be numerous, and it is desirable
to have them located at short intervals along every
airway. Municipal fields are included in the class of
permanent airdromes.

Airdromes: An ideal airdrome should be at
least 3000 feet square, so that it will afford ample
landing space for the largest planes. Every field
should have a straight runway at least 1800 feet long,
in every direction from which the wind is likely to blow.
This should be regarded as the minimum admissible
requirement, because if a motor should fail in leaving
a field having only an 1800-foot runway, there would be
danger of an accident in returning to the ground. A
3000-foot runway, on the other hand, would be long

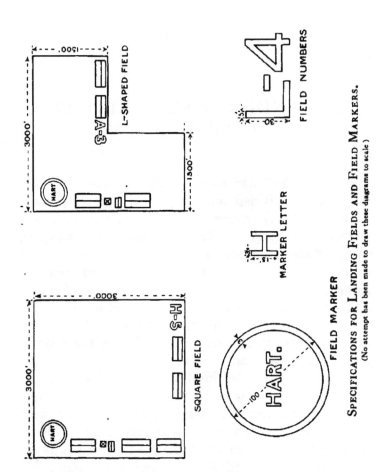

SPECIFICATIONS FOR LANDING FIELDS AND FIELD MARKERS.

(No attempt has been made to draw these diagrams to scale)

enough to permit the average pilot to take a machine off and still remain in a position to land without accident if the motor should fail in climbing.

The proper size of a flying field depends somewhat upon the nature of the surrounding country. If the land in the vicinity is cleared and suitable for emergency landings, the need of an airdrome 3000 feet square is not so urgent. If, however, the airdrome is surrounded by tall buildings, or if the adjacent territory is of such character that landing on it would be dangerous, safety considerations may demand an even larger amount of space.

In choosing the site for an airdrome, an effort should be made to find a place that can be easily reached, and one that is not likely to be soon surrounded by buildings. A site that can be expanded to meet future needs is highly desirable, and it should be so situated that transportation facilities are accessible. Electric power and a water supply should also be available.

The best shape for a landing field is a square, but a T-shaped or L-shaped field will suffice, provided it affords a runway of sufficient length. The ground should be hard and firm under all weather conditions,— a light, porous soil with adequate natural drainage being the most suitable. Clay soil invariably demands a special drainage system of tiling, to prevent the occurrence of unsatisfactory conditions in wet weather. The field should be smooth and level, and covered with sod.

Every landing field should bear a distinctive sign of some kind, easily recognizable from the air, and it is also desirable to have the name of the city marked on the ground. A white circle, 100 feet in diameter, has proved highly satisfactory for marking purposes. The line forming the perimeter of the circle, and the lines constituting the letters, should be three feet wide, and the letters should be about 15 feet square. Permanent markers can be economically made by laying out the desired design on the ground, digging out the soil to a depth of six inches, and filling the trench or excavated area with crushed stone coated with whitewash. Frequent applications of whitewash must be subsequently made, to keep the stone white and visible from a high altitude.

Since it is necessary for aviators to land and start against the wind, a large wind indicator in the form of a standard wind-cone, or a white cloth "landing T,"

Curtiss Aero Photo.

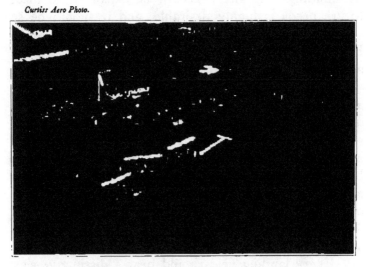

AN AIRDROME, SHOWING WIND INDICATORS.
(Note that the wind direction as indicated by the white T is not the same as that shown by the wind-cone.)

THE TYPE OF WIND-CONE USED AT CURTISS FIELD.

should be placed in one corner of the landing field. The wind-cone is considered the more reliable because it shifts automatically as the wind changes, whereas the T markers have to be moved by the workmen. Wind-cones are flown from poles, very much as flags are displayed. Hangars, repair shops, supply houses, emergency rooms, and gasoline stations should be located at the side of the field, in places where they will not interfere with the landing and taking off of the aircraft.

Provision should be made for night landings, and to this end large searchlights should be placed where they will throw a flood of light *on the ground* of the flying field. One powerful searchlight that will throw a beam of light straight upward should be placed among the buildings at the side of the field. A beam of this kind is visible from a high altitude on clear nights, and on foggy nights it illuminates the clouds and shows (when seen from above) as a brilliant white spot in the mist.

Every airdrome should be in charge of a superintendent, who should have a complete staff of fieldmen, mechanicians, riggers, and helpers. The superintendent's office should be located preferably at the side of the field, where it will be convenient to the shops and also to the landing area. An aviator should report to the superintendent of the field immediately upon landing, and state his wants. The superintendent will then assign the necessary men to take care of the machine, and to see that the visitor's needs receive proper attention.

Curtiss Aero Photo.

A WELL-ARRANGED AIRDROME.

Near the superintendent's office, the emergency and first-aid station should be established, where the surgeon can maintain his office and where the ambulance and emergency outfit can be located. A lookout should be on duty outside the emergency station, and should immediately notify the surgeon in the event of a crash, or a serious accident of any kind. In case of a minor accident the injured person should be treated by the surgeon in the first-aid room.

Emergency Fields: Emergency fields are, as the name implies, fields to be used only in cases of emergency or forced landing. They need be only about 1500 feet long in their greater dimension, to provide fairly safe landing facilities. All that is really necessary is to provide sufficient area, level and free from trees and other obstacles, to afford a comparatively safe landing place in case of emergency. Every such field should be provided with a wind-cone, however, to indicate the direction of the wind.

It is highly desirable that an emergency field be provided every ten miles along an airway. The average airplane can glide a distance equal to about seven times its altitude, so that a plane flying at a height of 5000 feet would never be out of gliding distance of a landing field, if emergency fields were situated ten miles apart along the route.

Airways and Air Routes: Air routes and airways have already been laid out to some extent, and a few aeronautic maps have been drawn up to assist the aviator in finding his way about the country. Airways have been established or proposed, extending across the continent from the Atlantic to the Pacific, and some have been laid out along the coast. Airways are transcontinental and coastal in extent, while shorter "air routes" interconnect the main airways.

Airways, as planned, are eighty miles in width and include within their boundaries the principal cities along the route. Airdromes, emergency fields, and aerial mail stations are rapidly being developed along the existing airways and are indicated on aeronautic maps by distinctive markings.

Aerial Laws: Aerial Law is as yet a new and largely undeveloped subject. The extent to which aerial navigation has developed necessitates the establishment of a strict code to be followed. The importance of such a code is realized and the problem is receiving serious consideration.

There is a great need of legal regulation of landing fields, both public and private, particularly with regard to properly restraining visitors and sightseers who come to the field for the purpose of looking on. Experience shows that at the present time visitors are permitted to roam about the field pretty much as they wish, and they are exposed to dangers (which they little appreciate) due to the landing and taking off of aircraft. A number of accidents have occurred that could have been avoided if certain areas had been set aside for flying, and certain other definite areas for the use of the public. It has also frequently happened that rescue work has been seriously hampered, after an accident, by the presence of a large crowd on the field. In one instance it became necessary for the men in charge of the field to use a stream from a fire-hose to keep the crowd away from a wreck sufficiently to allow the rescue party to liberate the injured aviators.

In addition to laws governing the establishment and

Courtesy "Motor Life." Copyright by Kadel and Herbert, N. Y.

AN AERIAL LIGHTHOUSE.

use of landing fields, there should be stringent and effective control of flights over centers of population. The hazard of flying, as it affects the person on the street, in his home, and in his workplace, is far greater than is generally understood. Laws should be passed, protecting the innocent population from exposure due to flights over cities, especially exhibition work and "stunting." Under ordinary circumstances, aircraft should be forbidden to pass over cities. Prohibited air channels have long been known in Europe.

The licensing of pilots is another factor that should be controlled by law. (See also Section IV, page 57.) The safest airplane in the world becomes unsafe and highly dangerous in the hands of a pilot who is not mentally and physically competent, and properly trained. Legal requirements should subject pilots to frequent examination as to their fitness, and licenses should be issued for a comparatively short period of time. At the end of this period the aviator should be subject to re-examination, as we have stated elsewhere. When these things are accomplished it can be said that a fair start has been made toward putting aerial navigation on a business basis.

Rules of the air, similar in nature to our marine laws, should be established and enforced. Aeronautic maps, similar to our present nautical charts, would provide accurate information for the air pilot, and a system of aerial lighthouses, marking the various air routes, would be of great assistance to aerial navigation at night. Conspicuous markers along the traveled air routes would serve the aeronaut in much the same way that a mariner is now assisted by the spars and buoys found in our navigable waters; and the systematic sending out of wireless directional signals from important airdromes would also tend strongly to insure safety.

VII. METEOROLOGICAL SERVICE

REALIZING the need for information regarding the weather conditions in the higher altitudes, the United States Weather Bureau, in co-operation with the various government flying fields, now furnishes data of this kind. Reports forecasting the weather conditions for the next twenty-four hours are prepared daily and sent to the various flying fields about the country.

In gathering the data from which these reports are made, kites, kite balloons, airplanes, and small rubber balloons, whose course through the air may be followed by the use of theodolites, are employed. The direction and velocity of the wind at various altitudes, together with the temperature of the air and the height of cloud formations, are noted. These records are made twice daily, and from them bulletins are prepared, giving the current and probable future atmospheric conditions at altitudes of 250, 500, 1000, 2000, 3000, and 4000 meters.

The country is divided into zones, and the flying fields in each zone receive the bulletins forecasting weather conditions for their vicinity. To aid pilots flying from one zone to another, special reports may be obtained upon request, covering the probable conditions over the entire route. Requests for such reports should include the name of the starting point, the destination, the route to be followed, and the time of contemplated departure.

Before starting a flight of any kind, a pilot should find out what conditions he is likely to meet aloft. The conditions on the ground afford but small indication of what is to be met with in the upper air, and no reliable information regarding the conditions at the higher altitudes can be obtained unless measurements and observations are taken. This work is also done for the aviator by the Weather Bureau, and he merely has to ask for the information, at the nearest weather station or flying field.

At certain of the meteorological stations, special upper air investigations are being carried on, and the activities at these stations have made it necessary to restrict the areas near by. During clear weather an aviator might see the kites and other kinds of apparatus in the air, but even under favorable atmospheric conditions they cannot be seen at any considerable distance, and in cloudy weather or in semi-darkness they are not visible at all; and if a plane should collide with the kites, or with the wires by means of which they are flown, serious damage would be likely to result.

Aviators are therefore warned not to fly above the stations at which work of this kind is being done. They are located at the following points: Broken Arrow, Oklahoma; Drexel, Nebraska; Due West, South Carolina; Ellendale, North Dakota; Groesbeck, Texas; and Royal Center, Indiana.

VIII. AIRCRAFT INSURANCE

LIMITED CHARACTER OF THE FIELD: A
consideration of aircraft insurance naturally
begins with an inquiry respecting the field for such
insurance. This field appears to be limited, at the
present time, to the heavier-than-air machines, and
the number of these that are employed in commercial
activities (as distinguished from exhibition and sport
purposes) is still quite small.

Lighter-than-air machines have not yet been
developed to a point where they are used to any extent
for private and commercial purposes, and they are
not considered to be within the field of aircraft insurance. The reasons for this are many, but prominent
among them are the difficulty and expense of equipping
and maintaining a lighter-than-air machine, and the
unreliability and grave danger attendant upon the use
of inflammable gas as a means of overcoming the
force of gravity. There is some promise that airships
buoyed up wholly or partly by gas will sooner or later
be developed to the point of practicability for private
and commercial use, and if they can be rescued from

their present unusual hazards, a great field of usefulness
can be found for them, because in many respects they
have an advantage over the other class of aircraft.

Speaking broadly, the field for aircraft insurance
is yet to be developed, because the aircraft themselves
are yet to be manufactured and sold.   The aircraft
now available for private and commercial use are largely
of the training type, and not fitted nor intended for
long-distance flights.   Strange as it may appear, the
large majority of the aircraft now available for private
ownership are much safer when used reasonably for
exhibition,—even including so-called "stunt flying,"—
than they would be for long distance flights, as a general
rule.   Of course manufacturers are developing new
types of airplanes, but at the present time the available
aircraft are chiefly used in enterprises that do not
furnish an attractive field for aircraft insurance and the
field at the present lies largely in the realm of conjecture.

The Development of Transportation: A railroad train, seventy-five or eighty years ago, was
more of a curiosity than an airplane is now.   In
those days small model trains, running on circular
tracks, were exhibited at county fairs and a fee was
charged to see the side show.   Railroads, however,
offer no opportunities for sport use, and their development was unhindered by any misapplication of their
purpose.   The railroad has become an absolute public
necessity and an important feature in our lives.

A number of years later, the trolley car was
developed.   In the early days a passenger often refused
to pay fare on one of these cars until it reached the
point at which he wished to alight, because too often
it did not reach that point.   Something was happening

all the time, and judging by the experiences of those days, it seemed that there would never be such a thing as a successful trolley system in the country.

Then came the automobile, and the history of its development is so recent that comment is hardly necessary. One incident of the early days of automobiles is recalled, in which a party undertook a run from Hartford, Connecticut, to New Britain, a distance of nine miles. A repair man and a car loaded with tools and appliances was taken along, to help out any cripples that might be picked up on the way. The repair car broke down, but fortunately the other cars reached their destination in safety.

For the first time since we had to depend upon the horse, we had found a means of land transportation that could also be devoted to sport purposes, and it promptly was so devoted, with the result that the development of the automobile was delayed by its misapplication. A speed craze took hold of the people. Automobile racing, exhibitions, and tricks became prominent in the early days of that means of transportation, and for a time these features promised to overshadow the really serious personal and commercial uses to which automobiles could be put. In its early days the automobile was cordially hated by a vast majority of the people, particularly in rural communities; but it survived and to-day it is found in almost countless numbers in all communities. It has come to be a necessity, and the people at large probably feel that they cannot dispense with its use.

Possibilities and Limitations of Aircraft: In the automobile as a means of transportation, we find something which more closely approximates the aircraft than any other present form of transportation, particularly by land. The all-important question from the insurance standpoint is, will the history of railroads, trolleys, and automobiles be repeated in the future history of aircraft? The automobile survived its purely personal and sport uses, and has developed into a commercial necessity. Will aircraft do the same thing? We can make no positive answer at present, but we can reason *toward* an answer by examining some of the claims that can be made in behalf of aircraft as a means of transportation.

Prominent among those claims is the matter of speed. The speed which has been obtained in airplanes is already phenomenal, and the possible speed of the future is beyond conjecture; but speed is not the only consideration. If a man, under urgent business requirements, can actually fly from New York to Chicago in eight or ten hours when transportation by train would require more than twice that time, that looks attractive on its face, and it looks as though airplanes might be developed as a means for the rapid transportation of passengers; but we must go a step further and consider the fact that a trip from New York to Chicago in a given number of hours is only a part of the story. The railroad stations in New York and Chicago are accessible, and when a passenger arrives at a railroad station there are convenient means of local transportation by the use of which the traveler can reach his actual destination speedily. It is not so with the airplane at present. If a man in the business center of Chicago desires to travel by airplane to New York, he must first journey to an outlying field which will necessarily be in the suburbs, and not necessarily within convenient reach by means of short local travel. Therefore, a fair portion of the time that is apparently saved by the airplane is spent in getting to a starting place. Then when he arrives at New York, the same situation is encountered. Perhaps he may land at Mineola or somewhere else on Long Island, and then have to use up an hour or more in traveling from that place to his actual destination in the business district of New York City. In this aspect the allurement of the airplane loses some of its force because the time actually saved, even if the trip be accomplished without mishap, is much less than it appears to be on the face of the record;—all of which goes to show that before the airplane can be recognized as a suitable and necessary means for rapid transportation, landing facilities must be provided with means of rapid transportation to business centers in the various cities of the country. Up to the present moment no substantial progress has been made in that direction, and all of this militates against the development of the airplane, and consequently against the development of the field for aircraft insurance.

The Dangers of Aerial Transportation: The dangers of airplane transportation have in the past deterred a great many people from accepting that means of travel, and will probably continue to deter them in the future. These dangers may be largely exaggerated. The railroads, the trolleys, and the automobiles have left behind them in the course of their development a long trail of dead and injured, and the fact that airplane transportation is dangerous probably will not of itself seriously delay the development of such transportation, if other obstacles are removed.

During the war the men who were trained in this country for aviation duties were trained upon fields that were more or less congested, and with machines that were more or less deficient. If available figures are correct, we must admit that even under the unusual conditions attendant upon war training, and the unusual hazards due to an undeveloped machine, there was only one fatality in nearly 3000 hours of flight; and 3000 hours of flight under favorable conditions and at a fairly moderate speed would take an aviator seven or eight times around the world, if such a thing were practically possible. There are reasons for believing that the frequency of fatal accidents in private and commercial use would be much lower than the frequency recorded during the period of war training, when all surrounding conditions are considered.

Figures compiled by the United States Air Service covering a period of six months during the war show that only about 2½ per cent. of all accidents, both fatal and non-fatal, were due to failure in the plane construction or its parts. The same tabulation shows that in the event of injury where a machine carries a pilot and one or more passengers, the pilot is the most likely to escape.

Making Insurance Rates: In the consideration of accident characteristics in airplanes for the purpose of reaching rate results, it has been assumed that the proportion of fatal and permanent total injuries to the total number of injuries would be far larger than it is in ordinary casualty lines. Whether this theory will prove true in practice or not remains to be seen; but in the absence of reliable data respecting this particular feature it seems reasonable to assume that in the distribution of accidents as to results, cases involving fatal and permanent total disability will occur in greater proportion than similar cases in other lines.

Another theory has been employed in developing compensation rates particularly, and that is, that in the event of fatal injuries the proportion of those found to be without dependents will be larger among airplane pilots than in the ordinary compensation lines. This is conjecture almost entirely.

The laws of many states requiring payment into a special fund in cases where there are no dependents will offset this conjecture to a considerable degree. So far as is known, there are no data of any moment which would serve to either prove or disprove this theory, but it has been used almost from necessity in order to produce an airplane rate (particularly for compensation) which was not on its face prohibitive, and which would not obstruct the progress of this new means of transportation.

Unreliability of the Airplane: The next feature which apparently weighs against the growth of the airplane as a means of transportation is the unreliability of the machine, as compared with other agencies that travel on the land or the sea. Perhaps the airplane does not differ much, in this respect, from the automobile in its earlier history, or even from the railroads or trolleys. Stability and dependability are matters of development, and if we have due faith in the inventive genius of our people, we may with reason conclude that present conditions in this respect will be materially improved, if not largely cured, in the near future. We may even now note that the alleged instability of the present airplane is not abundantly demonstrated by the evidence. During the two years or more that airplanes have been used for carrying mail, practically ninety-five per cent. of the contemplated trips have been successfully flown. This performance is regarded as unusually creditable, because during the winter months the flying was done under specially trying conditions, and many of the planes had to be equipped, for weeks, with snow skids in place of wheels.

The Cost of Airplanes: Airplanes, so far as they are available, can be purchased at almost any price that a person is willing to pay; but a heavier-than-air machine capable of carrying as much as a ton of merchandise (in addition to the pilot, crew, instruments, fuel, and other necessary sources of weight) would certainly be a large and expensive apparatus. We cannot say what the cost would be, but it would no doubt materially exceed $25,000. Here the lighter-than-air machine would have many advantages, because in construction (outside of the gas supply) it would probably be less expensive. Therefore, when the gas supply problem is settled (if it ever is) the field for the lighter-than-air machine will probably be found to consist in the transportation of dead weights, where sustained speed is not of any special importance.

We know very little at present about the cost of maintenance and repair, although we have a general understanding that it is pretty large. An engine used in an airplane is capable of perhaps 500 hours of service, although it is customarily removed and overhauled after a far shorter period.

Very little is known about the cost of fuel. Some

rather reckless statements have been made respecting fuel cost, as well as the cost of repairs and allowances for depreciation; but the sum total of the whole situation appears to be, that the initial cost, when combined with the cost of maintenance and use, is at the present time nearly, if not quite, prohibitive; and unless this obstacle can be reasonably reduced, the development of the airplane, for commercial purposes at least, will be slow. However, we are reminded of the apparently unanswerable objections that were advanced only a few years ago respecting the automobile, and we do not forget how successfully the claims respecting excessive cost, not only as to the original purchase price, but as to maintenance and use, have been dealt with. The economy of transportation by automobile truck has been adequately demonstrated, as is evidenced by the constantly growing use of these vehicles.

Why should Aircraft be Insured? The next question is, why should aircraft insurance be undertaken by casualty companies? Here an entirely different line of reasoning is encountered. It is well known that in spite of all delays and hindrances, a large number of airplanes are actually in operation in various parts of the country. These machines are mostly owned by the government, and operated either by the army or navy, or by the mail service; and operations of this kind obviously do not come within the field of insurance. After all, however, there are some airplanes left. There are such things in service as private and commercial airplanes. Their use in many instances involves the employment of pilots and of others who, in the course of their duties as employees, are required to fly; and in such cases the compensation

A Seaplane Crash.

(This photo was taken at the instant the plane struck the ground.)

After the Accident.

(This seaplane crashed on a beach filled with recreation seekers. The pilot and his two passengers were killed.)

laws in most of our states require insurance (or other satisfactory security) for the compensation obligation. In other words, if the owners of airplanes in civil life have employees, the law,—in a great many instances at least,—requires them to obtain insurance.   There are many reasons for claiming that insurance companies professing to write compensation lines would fail in their duty if they did not devise means for providing insurance in this line also, inasmuch as the law requires the employer to provide himself with it.   Therefore, perhaps the first reason why aircraft insurance has been undertaken is that it is the duty of insurance companies to provide it or to devise means through which it can be secured.   These considerations apply to workmen's compensation insurance only, but a company undertaking this line would naturally conclude that there should go with it such other lines as its corporate powers would allow it to write, and which would serve to increase premium receipts and to an extent improve the distribution in a necessarily limited field.   Included in that program would be public liability and property damage, as well as individual accident insurance for passengers and others exposed to the hazard of flying.

Nature of the Insurance Contracts: The public liability and property damage lines are in some respects similar to the corresponding lines now undertaken, by many casualty companies, in connection with automobiles.   They differ, however, in one essential particular, which is, that public liability policies do not ordinarily cover the passenger hazard.

The passenger hazard may be covered by a special endorsement attached to the public liability policy.

An Airplane that Fell in a City Street.

(Consider the danger that persons on the street were exposed to when this accident occurred.)

With the passenger hazard endorsement, the policy protects the owner of the airplane against suits arising from injuries sustained by the public while riding in the plane. The rates for this form of coverage are a percentage of the passenger earnings, with a minimum sum per passenger trip. The maximum limits of such a policy are at present $10,000/$30,000,—that is, the company will not accept a responsibility for more than $10,000 on any one person killed or injured in any one accident, nor more than a total of $30,000 in case several persons are involved in any one accident.

Individual accident insurance is issued in the form of a daily ticket policy, which becomes applicable at whatever hour the flight is started during a given day, and continues until four o'clock A. M. the following day. This ticket policy is for the principal sum of $5,000, with the usual indemnities for dismemberment and loss of sight, and also for disabilities of a temporary character. Weekly indemnities, however, apply only where the holder of the ticket is a man. Similar tickets, with the weekly indemnities eliminated, are issued to women. This ticket-policy plan has been developed so that a similar contract may be issued to cover a trip of any proposed duration,—including a round trip, if that is desired,—and contracts of this kind are called trip-ticket policies. The rates depend upon the length and character of the trip, and at present are largely matters of special negotiation in individual cases. There are also means provided for obtaining an annual personal accident policy, carrying a rider permitting flight in airplanes.

Life insurance with an aircraft permit may also be obtained. It is issued in one-year non-renewable term form only, and an extra premium in addition to the term rate is charged.

It really matters little if the field in the near future is to be restricted, or if the final development of aircraft stops at a point short of making it a necessity in our personal and commercial lines. The airplane has come to stay, beyond any question, and its future depends largely upon the preparation made for its acceptance and regulation.

The Future of Aerial Navigation: In the foregoing pages, the requirements for successful aerial navigation have been discussed at length, and they are of utmost importance. We may assume that airplanes will be improved, and that they will be rendered more stable, more dependable, and less difficult of operation. Perhaps they will be much less expensive in original cost and subsequent upkeep in the near future, but all this will be of little avail unless aerial navigation in all its phases is made the subject of constant, careful, rigid supervision, under the operation of well devised and fully enforced laws.

Our former allies, as well as our former enemies, are far ahead of us in the development of aircraft, and in the development of the necessary insurance plans to go with it. It is freely stated that England is to be the aircraft center of the world, and that from England will come the insurance plans and provisions in the various lines required, and without which aircraft projects of whatever nature cannot succeed. The United States ought to strive to at least divide this honor with her former ally. We ought to make it our business to see that the United States remains on the map in the matter of aircraft development. The heavier-than-air machine was born in the United States, and developed here to a certain point, but it was ignored as a factor of any value by the people at large. Other countries took up the projects which we neglected, and we have suffered much in consequence. If reconstruction is ever accomplished,—if we ever come back to normal times and to a normal method of living,—there will come a period of sharp competition during which recourse must be had to every possible method for maintaining our position in our own markets and in the markets of the world. In this period aircraft will most surely play its part, and a very important part too. The fact that we here in the United States are far behind England and all other countries in the development of this most helpful competitive means should not deter us from laying a sound foundation and establishing a useful practice for aircraft insurance,—notwithstanding present discouragements, notwithstanding a limited field, and notwithstanding the lack of substantial hope for the immediate future.

We must have aircraft. They must be developed and improved. They must be cheapened in cost and up-keep. They must be dependable. They must be practical. They must do their part in the annihilation of space,—a part now so amply played in the transmission of words by telegraph, either of the old kind or of the more modern sort. Chicago can order goods from New York by telegraph or telephone in a few minutes, but New York cannot deliver the goods to Chicago, by the means now at hand, with sufficient promptness to meet the requirements of the future. The railroads are largely impotent. Steamships are

not always available. The automobile truck has its
limitations, and certainly aircraft has a place. We
must not forget these things. We must work in antici-
pation of the future. We must work for the supremacy
of the United States of America in all things respecting
her commerce, and those plans for commerce which
experience has demonstrated to be feasible, and which
probably will soon be regarded as necessary.

GLOSSARY OF AVIATION TERMS

*I*N *preparing this section we have endeavored to conform as closely as practicable to the terminology given in Report No. 91 of the National Advisory Committee for Aeronautics,—"Nomenclature for Aeronautics,"—and in a few cases we have adopted the precise language of that report. The present glossary is far more limited in scope than Report No. 91, however, because we have compiled it with special reference to the probable wants of readers of this book.*

Aerodrome: See AIRDROME.

Aerofoil: A flat or curved winglike structure, so designed and disposed that the air through which it moves will react against its surface. The part of an airplane wing against which the supporting action of the air is exerted. Often called a "plane." (The last syllable, "foil", refers to the fact that in the earlier forms, at all events, the aerofoils were quite thin.)

Aeroplane: See AIRPLANE.

Aileron: A movable auxiliary surface, usually constituting part of the trailing edge of a wing to which it is attached by a hinge, and used for controlling the rolling motion of an airplane. (A French word, meaning "a small wing or fin.")

Aircraft: Any form of vehicle used in navigating the air.

Airdrome: A large, permanent flying field,—usually equipped with hangars, repair shops, and a supply station. The word "aerodrome," from which "airdrome" is derived, was originally used by Langley to designate any heavier-than-air flying machine (of the airplane type), which is propelled by its own motive power; but in that sense it has become wholly obsolete. ("Aerodrome" is derived from two Greek words meaning "air" and "running.")

Airplane: A form of aircraft, heavier than air, employing wing surfaces for support, and containing a power plant for propelling it through the air. (Usually employed in connection with machines fitted with landing gear suited to operation from the land.)

Airscrew: See PROPELLER.

Airship: An elongated lighter-than-air machine, propelled by airscrews, and depending for its buoyancy upon a large bag or other receptacle, which is filled with a gas lighter than air. (Special forms of it are known as "dirigibles" and "blimps," and by other names.)

Air-speed indicator: An instrument for measuring the velocity of an aircraft relatively to the air through which it is moving.

Alcohol: This word, when used without any qualifying adjective, is supposed to signify ethyl alcohol, or grain alcohol.
(Compare METHANOL.)

Altimeter: An instrument for indicating the height of an aircraft above the surface of the earth. (Essentially an aneroid barometer.)

Anemo-meter: An instrument used for measuring the velocity of the wind, relatively to the body to which the anemometer is attached.

Angle of attack: The acute angle between the direction of the relative wind and the chord of an aerofoil.

Angle of incidence: The angle that a chord of an aerofoil makes with the horizontal, when the machine is in a flying position.

Aspect ratio: The ratio between the spread and the mean chord of an aerofoil.

Aviator: Any person (of either sex) who practises the art of flying in heavier-than-air machines.

Axes of an aircraft: Certain imaginary lines of reference, fixed with respect to the aircraft, and passing through its center of gravity. They include (1) the longitudinal or fore-and-aft axis, lying in the plane of symmetry and usually running in a direction parallel to the axis of the propeller; (2) the normal axis (sometimes called the vertical axis), also lying in the plane of symmetry but running in a direction perpendicular to the longitudinal axis; and (3) the lateral (or athwartship) axis, running perpendicularly to the plane of symmetry and intersecting the other two axes at the center of gravity of the machine.

Back-wash: A stream of disturbed air in the wake of an airplane. This is also referred to as the "slipstream."

Bank: To rotate an airplane through a limited angle about its fore-and-aft axis, so that one wing becomes lower than the other. To "right bank" is to incline the *right wing* downward.

Barograph: An instrument used for recording barometric pressures or altitudes.

Bay: The cubic space lying between two transversely adjacent pairs of struts in the truss of an airplane wing. The bay nearest the center of the machine is known as the *first* bay.

Biplane: A form of airplane having two sets of supporting surfaces, one above the other.

Blade: The paddle-shaped portion of a propeller, outside of the boss or hub. The working surface, or side against which the air thrust operates (and which is nearly flat), is the *face* of the blade, and the opposite, strongly-convex side is the *back*.

Blinker: See Non-skid.

Body: In an airplane or seaplane, the boat-like portion that carries the passengers, pilot, freight, etc. (Compare Fuselage and Nacelle.)

Boss: The hub, or central portion, of a propeller screw.

Camber: The rise or convexity of the curve of an aerofoil from its chord, usually expressed as the ratio between the maximum departure of the curve from the chord and the length of the chord. (This term has long been used in a similar sense in connection with bridge construction.)

Capacity: Same as LOAD.

Carrying capacity: See LOAD.

Ceiling: The *absolute ceiling* is the limiting altitude (measured from sea-level), above which a given aircraft is incapable of maintaining flight. The *service ceiling* is the height, similarly measured, above which a given aircraft cannot rise at a rate exceeding a certain small given limit,—this specified limit being 100 feet per minute in the United States Air Service.

Center of pressure: The point on any given aerofoil or aerofoil chord, through which, at any instant, the line of action of the resultant air pressure passes.

Chassis: See LANDING GEAR.

Chord of an aerofoil section: A straight line, parallel to the central plane of symmetry of the machine, and tangent at the front and rear to the under curve of an aerofoil section. The "length of the chord" is the width of the aerofoil as projected on the chord. (If the aerofoil has a doubly convex camber, the chord is understood to be the straight line joining the leading and trailing edges, and the length of the chord is then taken to be the distance between these two edges.)

Control column: A control lever with a rotatable wheel mounted at its upper end. The elevators are operated by the fore and aft movement of the column and the ailerons are actuated by rotating the wheel.

Control-stick: The vertical lever by which certain of the principal controls in an airplane are operated. This is sometimes called the "joy-stick."

Controls: A general term applied to the apparatus provided to enable the pilot to regulate the speed, direction of flight, attitude, altitude, and power of an aircraft.

Critical angle: The angle of attack at which the flow of air about an aerofoil changes abruptly. An aerofoil may have two or more such critical angles, and one of them usually corresponds to the position at which the lift of the airplane, in sustained flight, is greatest.

Decalage: The angle between the chords of the principal planes and the horizontal tailplanes. (This is also known as the "Longitudinal V.")

Dihedral angle: The main supporting surfaces of an airplane are said to have a dihedral angle when both right and left wings are upwardly or downwardly inclined to a horizontal transverse line. The angle is measured by the inclination of each wing to the horizontal. If the inclination is upward, the angle is said to be positive; if downward, negative. The several main supporting surfaces of an airplane may have different amounts of dihedral.

Diving rudder: Same as ELEVATOR.

Dope: The special varnish used for coating the cloth surfaces of airplane members, to render them taut, airtight, and waterproof.

Drag: The component, parallel to the relative wind, of the total force on an aerofoil or aircraft due to the air through which it moves. That part of the drag due to the wings is called "wing resistance," and that due to the rest of the airplane is called "structural" or "parasite resistance."

Drag wires: Cables used to prevent the horizontal pressure of the wind from folding the wings of an airplane backward. These are also known as drift wires.

Drift: The angular deviation from a set course over the earth, due to cross currents of wind.

Drift wires: See DRAG WIRES.

Elevator: A hinged auxiliary surface, usually attached to the tail plane, and used for controlling the attitude of an aircraft with respect to its athwartship axis,—that is, for imparting a *pitching* motion to the machine, or for counteracting such motion.

Emergency field:
A small landing field, for use in case of a forced landing. (Compare Airdrome.)

Empennage:
Same as Tail. (Derived from the French word "penne," a feather, in reference to the tail-feathers of a bird.)

Engine, right or left-handed:
The distinction depends upon the direction of rotation of the output shaft. If, when viewed from the output shaft end, the shaft rotates counter-clockwise, the engine is said to be right-handed. If the rotation is clockwise when the shaft is thus viewed, the engine is left-handed.

Engine bearers:
The rails, beams, or fuselage members which bear the weight of the engine, and to which the engine is bolted.

Entering edge:
The foremost edge of an aerofoil.

Factor of safety:
The ratio that the ultimate strength of any part bears to the maximum stress to which that part may be subjected in the course of normal operation. Thus if the stress that would be required in order to produce rupture is ten times as great as the maximum working stress, the part is said to have a factor of safety of ten.

Fins:
Small fixed plane surfaces, attached to various parts of aircraft for the purpose of promoting stability, and projecting in a way suggestive of the fins of a fish.

Flight path:
The path of the center of gravity of an aircraft, with reference to the surface of the earth.

Float (or Pontoon):
That portion of a seaplane which provides buoyancy when the machine is resting on the surface of the water.

Flying boat:
A seaplane having a boat-shaped body which serves as a float when the machine is resting on the

water. (Flying boats are often provided with additional auxiliary floats or pontoons.)

Foot bar: A pivoted bar, by which the pilot operates the rudder of his machine.

Fuselage: The elongated boat-like housing of an airplane, which contains the passengers and usually considerable of the mechanism also. It is approximately streamlined in form, and takes its name from the French word "fuseler," meaning "to taper."

Gap: The shortest distance between the planes of the chords of the upper and lower wings of a biplane, measured along a line perpendicular to the chord of the upper wing at any designated point of its entering edge.

Glide: To descend at a normal angle of attack without the aid of the engine,—the necessary flying speed being maintained by the action of gravity. (Gliding is often called *volplaning*.)

Gliding angle: The angle that the flight path makes with the horizontal, when gliding toward the earth under the influence of gravity alone.

Guy: A rope, chain, cable, wire, or rod, attached to an object to guide or steady it, or to hold it in position.

Hangar: A shed for housing aircraft. (A French word signifying a shed, shanty, or lean-to.)

Head resistance: The total resistance (in the direction of the longitudinal axis of the machine) that is offered by the air to the normal forward motion of an aircraft and its passengers or other contents.

Helicopter: A form of aircraft deriving its lifting or sustaining power from the direct vertical thrust of large downwardly-directed propellers. Machines of this type are not in actual use, but some authorities consider that the idea is capable of practical development. (The name is derived from two

Greek words, signifying "spirally-acting-wing.")

Horn: A short projection, often horn-like in form, secured to a movable part of an airplane and serving as a lever arm in controlling that part.

Hydro-airplane: A seaplane fitted with a landing gear of pontoons.

Inclino-meter: An instrument for measuring the angle between any axis of an aircraft and the horizontal.

Joy-stick: A colloquial expression used to designate the control-stick.

Land plane: An airplane having landing gear for operating from the land. (Compare SEAPLANE.)

Landing gear: The understructure of an aircraft, including all those parts which are designed to support the machine when it is not in flight, and to enable it to alight in safety when making a landing. (Also known as the "chassis,"—a French word meaning a "framework" or "housing.")

Lateral stability: Stability with reference to rotation about the longitudinal axis.

Leading edge: Same as ENTERING EDGE.

Lift: The component of the total air force that is perpendicular to the relative wind and in the plane of symmetry.

Lifting capacity: Same as LOAD.

Limiting height: See CEILING.

Load: The total maximum weight that an aircraft is capable of supporting in flight. The *useful load* (sometimes called the "carrying capacity") is the maximum weight that the machine can support in addition to its own weight and that of the various instruments that are essential to its proper management.

Longeron: A fore-and-aft member of the body-frame or float (Also called a "longitudinal.")

Longitudinal:	See LONGERON.
Longitudinal stability:	Stability with respect to pitching,—that is, with respect to rotation about the athwartship axis. (See STABILITY, and AXES OF AN AIRCRAFT.)
Methanol:	The substance commonly known as methyl alcohol or *wood alcohol*. (The American Chemical Society has recommended the adoption of this word, to avoid confusion with grain alcohol; and we have followed that counsel in the present book.)
Monoplane:	A form of airplane deriving its support from a single wing on each side of the body.
Multiplane:	A form of airplane employing more than two sets of main supporting surfaces, superposed. (Compare BIPLANE and MONOPLANE.)
Nacelle:	The body portion of a pusher-type airplane, or, the car of a dirigible. (In French, the word means "a little boat.")
Non-skid:	An auxiliary vertical plane, sometimes placed upon or between the wings of an airplane to check side-slipping. (Also called a "blinker.")
Nose dive:	A dangerously steep head-on descent.
Ornithopter:	A form of aircraft deriving its propelling force and support from wings that flap similarly to those of a bird. Machines of this type are not yet practicable. (The name is derived from two Greek words signifying "bird wings.")
Pancake:	To descend vertically, or along a very steep course, with the wings of the machine nearly horizontal. (The name bears reference to a fancied analogy with the flatwise fall of a pancake.)
Parachute:	An umbrella-like appliance used to retard the descent of a falling body, and particularly of a person. (From two Greek words signifying "to ward off a fall.")
Pilot:	A person qualified to operate an aircraft.

Pitch: The *geometrical pitch* of an airscrew or propeller is the distance it would travel forward, in one revolution, if it were turning in a medium that would allow of no slipping. The *effective pitch* is the distance the airplane advances, in actual flight, in the course of one revolution of the propeller. The difference between the geometrical pitch and the effective pitch is called the *slip* of the propeller.

Plane: See AEROFOIL. (The word "plane" is often used, also, as an abbreviation for "airplane" or "volplane.")

Plywood: A structural material made by gluing together a number of layers of wood veneer, with the grain of the several layers running in different directions.

Pontoon: See FLOAT.

Propeller: A screw so mounted on an airplane that its rotation moves the machine through the air.

Pusher airplane: An airplane in which the propeller is located back of the wings, so that it pushes the machine from behind. (Compare TRACTOR AIRPLANE.)

Pylon: A mast or post.

Race of an airscrew: The air-stream delivered by a revolving propeller.

Relative wind: The motion of the air relatively to the airplane. The direction and speed of the relative wind depend (1) upon the direction and speed of the actual wind as perceived by a stationary observer, and (2) upon the direction and speed of the motion of the airplane itself.

Resistance: See HEAD RESISTANCE.

Rib: A fore-and-aft member of the wing, used to give form to the wing-section, and to serve as a support for the wing covering.

Roll: To incline an airplane laterally. The angle of roll is the angle through which an aircraft must rotate on its longitudinal axis in order to bring its lateral axis into the horizontal plane.

Rudder: A flat or streamlined surface, hinged or pivoted, and used for controlling the movement of an aircraft about its vertical axis.

Seaplane: A special form of airplane, designed for operating from the surface of the water. (The term "water plane" is passing out of use, the word "seaplane" being preferred, whether the machine is to operate from the ocean or from rivers and lakes.)

Side slip: A sidewise and downward motion of an airplane, taking place at right angles to the normal direction of flight, in such a way that the wings of the airplane move through the air edgewise or nearly so.

Side slipping: Sliding *toward* the center of a turn, on account of banking too steeply, or in consequence of having too low a flying speed. (Compare SKIDDING.)

Skidding: Sliding sidewise and *away from* the center of a turn, on account of insufficient banking. (Compare SIDE SLIPPING.)

Skids: Long metal or wooden runners fitted to an airplane, and designed to prevent nosing-over when landing, or to protect the wings from contact with the ground, or to support the tail of the machine. (Also applied to a ski-form of landing gear, which is used on snow.)

Skis: Same as SKIDS.

Slip: See PITCH; SIDE SLIP; SIDE SLIPPING; SKIDDING.

Slip-stream: Same as BACK-WASH.

Spar: See WING SPAR.

Spin: An aerial manoeuver consisting of a combination of roll and yaw, with the longitudinal axis of the airplane inclined steeply downward. The airplane descends in a helix of large pitch and small radius, the upper side being inside of the helix and the angle of attack on the inner wing being maintained at an extremely large value.

Spread: The maximum width of an airplane, as measured from tip to tip of the wings.

Stability: An airplane is said to possess *stability* if, after being subjected to a small disturbance of any kind during steady flight, it tends to return quickly and automatically to a similar steady state. Various kinds of stability are recognized, according to the nature of the disturbance. For example, the machine may or may not be stable with regard to rolling, yawing, pitching, or motions of other specified kinds; and it may be stable with respect to a disturbance of one of these kinds, but unstable with regard to another one.

Stabilizer: Specifically and most commonly, the stationary horizontal tail surface of an airplane. The name is also applied, however, to any mechanical device the purpose of which is to insure, or increase, stability in flight.

Stagger: The amount by which the entering edge of the upper wing of a biplane, triplane, or multiplane projects beyond the entering edge of a lower wing, —usually expressed as percentage of gap.

Stagger wires: Wires used for maintaining the stagger of the wings.

Stalling: The losing of the speed (relatively to the air) that is essential to the proper control of an aircraft.

Statoscope: An instrument for detecting or registering small changes in altitude.

Stay: A wire, cable, rope, or rod, used for holding parts together, or for contributing stiffness.

Step: A step-like break or discontinuity of form on the bottom of a float or hull, designed to modify the dynamic reaction from the water.

Stick: Same as CONTROL-STICK.

Streamline: The path followed by any given particle of air, in a current that is flowing (or streaming) around a solid object. The object may be fixed and the air moving, or the object may be moving while the air is stationary (save for the local disturbance that the object produces). In either case the streamlines are supposed to be the lines of flow of the air, as they appear to an observer who is fixedly associated with the solid object. The word "streamline" is usually understood to refer to streaming motions that are not attended by the production of eddies. Eddy-formation involves the expenditure of energy, and as the energy so expended is wholly wasted, it is desirable to give every part of the airplane a streamline form (or section) so that the air will flow around it without producing eddies.

Strut: A compression member of a truss frame; specifically and most commonly, one of the vertical members of a wing truss in a biplane.

Sweep-back: The angle between the lateral axis of an airplane and the entering edge of the main plane, measured in a plane parallel to the lateral axis, and to the chord of the main plane.

Tail: The rear portion of an airplane, including the rudder, elevators, and fins.

Tail plane: A stationary horizontal (or nearly horizontal) tail surface, used to stabilize the plane with regard to pitching motions. This surface is often called a "stabilizer."

Tandem plane: An airplane having two sets of wings, one behind the other.

Torque of a propeller: The torsion to which the propeller shaft is subjected when the engine is in motion; the tendency of the propeller to cause an airplane to revolve about its longitudinal axis, in a direction opposite to that in which the propeller is turning.

Tractor airplane: An airplane having its propeller situated in front of the wings, so that the machine is *pulled* through the air. The propeller of such an airplane is often called a "tractor screw" or "tractor airscrew." (Compare PUSHER AIRPLANE.)

Trailing edge: The rearmost portion of an aerofoil.

Triplane: An airplane having three pairs of main supporting surfaces or wings, superposed.

Truss: The framework by which the load sustained by the wings of an airplane is transmitted to the body of the machine.

Under carriage: Same as LANDING GEAR.

Useful load: See LOAD.

Volpique: Same as NOSE DIVE. (From two French words, "vol" and "pique", signifying, respectively, "flight" and "beak" or "prow.")

Volplane: Same as GLIDE.

Warp: To change the form of a wing by bending or twisting it, by means of controls provided for that purpose.

Water plane: See SEAPLANE.

Wing: That part of the main supporting surface of an airplane which lies on one side of the body or fuselage. Thus a monoplane has *two* wings, and a biplane has *four*.

Wing loading: The weight carried per unit area of supporting surface.

Wing rib: Same as RIB.

Wing spar: A stiffening chord or member in a wing, running athwartship;—*i. e.*, lengthwise of the wing.

Wood alcohol: Called "methanol" in the present book. (See METHANOL.)

Yaw: To swing an airplane about its vertical axis.

The "angle of yaw" is the angle between the direction of the relative wind and the plane of symmetry of an aircraft. It is positive when the aircraft turns to the right.

Zoom: To climb for a few moments at a steeper angle than can be maintained in continuous flight.

INDEX

CPSIA information can be obtained
at www.ICGtesting.com
Printed in the USA
LVHW081029120123
737030LV00018B/166

9 781018 263670